Research Ethics for Scientists

Research Ethics for Scientists

A Companion for Students

C. Neal Stewart Jr.
University of Tennessee

WILEY-BLACKWELL
A John Wiley & Sons, Ltd., Publication

This edition first published 2011

Wiley-Blackwell is an imprint of John Wiley & Sons, formed by the merger of Wiley's global Scientific, Technical and Medical business with Blackwell Publishing.

Registered office:
John Wiley & Sons, Ltd, The Atrium, Southern Gate, Chichester, West Sussex, PO19 8SQ, UK

Editorial offices: 9600 Garsington Road, Oxford, OX4 2DQ, UK
The Atrium, Southern Gate, Chichester, West Sussex, PO19 8SQ, UK
111 River Street, Hoboken, NJ 07030-5774, USA

For details of our global editorial offices, for customer services and for information about how to apply for permission to reuse the copyright material in this book please see our website at www.wiley.com/wiley-blackwell.

Library of Congress Cataloging-in-Publication Data

Stewart, C. Neal Jr.
Research ethics for scientists : a companion for students / Neal Stewart.
p. cm.
Includes bibliographical references and index.
ISBN 978-0-470-74564-9 (pbk.)
1. Research–Moral and ethical aspects. 2. Scientists–Professional ethics. I. Title.
Q180.55.M67S76 2011
174′.95–dc23

2011016038

A catalogue record for this book is available from the British Library.

This book is published in the following electronic formats: ePDF 9781119978879; Wiley Online Library 9781119978862; ePub 9781119979869; Mobi 9781119979876

Typeset in 10/13pt Bembo by Aptara Inc., New Delhi, India
Printed and bound in Singapore by Fabulous Printers Pte Ltd

First Impression 2011

Contents

Preface

My initial involvement with research ethics was quite accidental (to me) and commenced just as I began my own PhD programme as a student. I was selected by the Associate Dean of the graduate school to be the Chief Justice of my university's graduate honour system. To this day, I still don't understand how that all happened, but now I realise the huge affect it subsequently had on my career. Unbeknownst to me at that time, it paved the way for this book some 20 years later. As Chief Justice, my duties were to help investigate and hear cases of plagiarism, research misconduct, and cheating in courses by graduate students – my peers. I still recall my major professor's response when I asked him what he thought about my taking the job. "If you don't mind judging your fellow students..." In other words, I don't think he believed it was such a good idea. I wasn't altogether convinced about this new gig either – I thought it had the potential to be a significant diversion from the research I needed to do to graduate. Plus, truly, what scientist wants to judge the allegedly bad practices of his fellow peers in research? This, I find, is a common feeling among scientists. Few scientists are comfortable policing the conduct of other scientists.

The Graduate Honour System cases of alleged student misconduct were heard and decided by a panel of faculty members and graduate students. I simply presided over the proceedings and administered the system. If a guilty verdict was found, then a penalty would be prescribed, and I was the guy to tell the accused of their fates. These penalties ranged from probation to dismissal. After the hearings I walked downstairs from the hearing room and into the ersatz waiting room, personally delivered the good or bad news to the graduate student; always an anxious moment. This simple bearing of good or bad news showed me in a profound way that there is a face and heart behind every case of scientific misconduct.

Hearing these cases over three years opened my eyes to the world of bad behaviour in science (and most of the cases we heard were in fields of science or engineering) that I hadn't realised even existed. It also helped me understand some of the psychology and pressures that precipitated academic misconduct. That experience helped steer my own career clear of major potholes and fatal wrecks alike. Oh, I still made my share of mistakes, but none were fatal. I had simply been given the somewhat unique chance to learn from lots of other people's mistakes. And I think I could have steered clear of a few more of my wanderings had I read a book such as this one and/or sat through a one-hour graduate course on research ethics. I'll make my own confessions throughout the book, and we will examine real and fictional case studies that should be fuel for thought as scientists wind their way through their careers.

With my PhD in hand and the busy day-to-day tasks of running a lab and teaching, the days of my ethical "trials" were a distant memory. Real-life research integrity didn't hit home until just a few years ago when I was the "victim" of plagiarism. I vividly recall reading my own words from another person's paper and thinking, "this looks familiar – and the writing's not so hot." A student's plagiarism of my own work inspired me to pursue ethics anew in the form of co-teaching a graduate course on practical research integrity. This book then naturally arose from my teaching experiences, and from the fact that when my colleague and I searched for a book or material to help teach our graduate-level research ethics course, we learned there are a plethora of works on bioethics and many fewer that address research ethics. As a practicing biologist, I don't consider this book to be a scholarly treatise in ethics; it is written to practically address common problem issues in scientific research with narrative and case studies. I wrote it as a guidebook of sorts – both for undergraduate students contemplating a life in science and those graduate students and early career scientists who find themselves in the thick of it. In the end, the book turned out to be more autobiographical than I'd set out for it to be. That said, all opinions are my own and all names I use in the fabricated case studies are also fabricated. Any resemblance to real people is purely accidental.

I am thoroughly convinced that the best ethical practices lead to the best science. Granting agencies such as the National Institutes of Health (NIH) and the National Science Foundation in the US must agree as they require research integrity training to their awardees. I think it is simply a matter of time before all US funding agencies follow suit. I see more and more scientists now motivated to teach courses in research ethics to address these needs. Aside from mandates set by funding agencies, there seems to be a growing number of colloquia, informal meetings and workshops on research ethics being held. This is a welcome trend to proactively address real concerns in a complicated research world. Research integrity is for everybody!

Knoxville, TN, USA
March 2011

Acknowledgements and Dedication

Many people have shaped my life and career and have therefore shaped this book. I'm greatly indebted to my scientific mentors who took a chance on me as a trainee: Erik Nilsen and Wayne Parrott. For each I was unproven and a significant risk, but they saw past the risks to the potential. They are both superb mentors. I'm also indebted to my own trainees. My career was born and sustained because of these tireless and dedicated individuals who work the pipettors so I can attempt to make contributions in other ways. In addition, they and others have taught me innumerable and valuable lessons about best practices in science research.

I'm grateful to people who have joined me in teaching research ethics, especially Lannett Edwards who co-founded our course four years ago. Charlie Kwit and Lana Zivanovic joined me in teaching research ethics last year and graduate students H.S. Moon and Blake Joyce were teaching assistants and acted as peer teachers in the 2009 version of teaching ethics. In 2010, Mark Radosevich joined Lana in teaching the course and graduate students Charleson Poovaiah and Jonathan Willis have acted as teaching assistants. Without EPSN funding for partial graduate teaching assistantships for these four students, we'd all have been poorer without their input in our course. I'm also grateful for the help provided by graduate student Kim Nagle during the class. I've learned a lot about ethics from teaching with all of these capable individuals.

I include also Gary Comstock in this list of key people to thank. When I first got interested in teaching research ethics I was fortunate enough to call Gary to get his advice on the subject and attend one of his research ethics workshops. His vision and input was critical to what the course, and ultimately this book, became. He is a real professional in this field and is one of its leaders.

So many people helped on the book manuscript by rendering figures, organisation, proofreading, editing, and getting permissions, among other things. At the top of this list are our group's able administrative specialists, Michelle Hassler and Jennifer Young Hines, who did much of the administrative work (and there was lots of it!) for the book. Reggie Millwood, Blake Joyce, H.S. Moon, Mitra Mazarei, Virginia Sykes, Dave Mann, Muthukumar Balasubramaniam, Jonathan Willis, Jason Abercrombie and other people in my lab played critical roles in contributing and fine tuning the contributions.

Thanks to Bob Langer and Daniel Anderson for allowing me to interview them on mentorship. Bob, especially, has personally inspired me to become a better mentor. Unbeknownst to him, he was also the inspiration for me to allow my lab to continue its growth from my self-imposed and arbitrary cap.

Thanks to Izzy Canning, Fiona Woods, Rachel Wade, and all the great people at Wiley-Blackwell in Chichester for the encouragement and guidance throughout this project. They were both kind and firm in their guidance, i.e., the perfect editorial staff. I also sincerely appreciate the time that the many peer-reviewers took to critique the various stages on the manuscript. I typically did not look forward to receiving the reviews, and then was not very happy with much of what they suggested, especially in the early stages, but in retrospect, their advice was critically important to the quality of the book. I owe a debt of gratitude to both the editorial team and reviewers.

This book is dedicated to my first and best mentor, Charles Neal Stewart, Sr. (1930–2010), who looked forward to seeing this book in print. I talked about this project with him on a regular basis during its development and he encouraged me to see it to completion. Thanks Dad.

To God be the glory.

Chapter 1

Research Ethics: The Best Ethical Practices Produce the Best Science

ABOUT THIS CHAPTER

- Research science is becoming increasingly complex and riddled with pitfalls and temptations.
- Global competition and cooperation will likely change the face of science in the future.
- Science is an iterative loop of ideas, funding, data, publication, in turn, leading back to more ideas.
- Ethics can be a guide toward best practices.
- Best scientific practices lead to the best science results and discoveries.
- Best practices and mentorship produce the best scientists.

It seems that it is increasingly difficult to be a research scientist. The number and complexity of rules, electronic forms, journals and publishing, and government and university regulations are ever-growing. The competition for funding is often ruthless, and the criteria exacted to warrant publication in good journals also seem to be on the rise. Indeed, not just the pressure to publish, but the pressure to publish the "right" papers in the "right" journals is also increasing. Nominally, the preparation of proposals and publications has been ostensibly made simpler by computer technology, yet the potential for real- and faux-research productivity has also been enabled by computers. Technology is a double-edged sword: enabling high levels of knowledge creation and dissemination, but also enabling research fraud and shoddy science. Thus, ethical dilemmas seem to be appearing at an increasingly rapid pace, with research misconduct regularly being the subject of news articles in *Science, Nature*, and *The Scientist*. I wouldn't be surprised when and if these scientific periodicals hire ethics reporters who will specialise in reporting misbehaviour. Even people who don't keep up with science news are familiar with the term "cold fusion" and the infamous stem cell cloning and data fabrication case from South Korea. While the most notorious cases of misconduct have occurred in higher-profile fields of science, such as physics and biomedicine, it is clear that no area of science is immune to unethical behaviour (Angell 2001; Judson 2004).

Research Ethics for Scientists: A Companion for Students, First Edition. C. Neal Stewart Jr.

We live in a "multiscience" world. Multitasking, multidisciplinary work and multi-authored works, to name a few, are ingrained in the fabric of science culture and certainly multi-multi is expected in order to succeed and move up the scientific ranks. The isolated small laboratory with the lone professor and few staff (see Weaver 1948 for a perspective) has given way to larger labs interacting in complex collaborations in interdisciplinary science. Complex relationships are accompanied with tough decisions regarding authorship, dicing the funding pie, and how to treat privileged data. And immense amounts of data at that, which are shared (or not) and curated in useful and meaningful ways (or not). In all this mix, the temptation to cheat, cut corners, and misbehave seems to be at its zenith for scientists wishing to compete at the highest levels of science, striving to get tenure and become rich and famous. Of course, one alternative to honest competition and competence, as seems to be the case for some scientists, is to con their way to the top. Cheating is front page news in business, politics and sports sections alike. Perhaps a bigger problem to outright fraud is cutting ethical corners. Thus, we have an apparent paradox – the antithesis of this chapter title – that the best (or highly rewarded) science is compromised with seemingly endless ethical issues. Whereas the lone professor and his or her graduate student worked in simpler and more linear paths in the past, modern science seems far too convoluted for its own good (Munck 1997). How can we win? How can sound science prevail in the face of all the obstacles?

If the situation is not complicated enough, it seems that there is growing concern about the abuse of graduate students and postdocs by their mentors. Some senior scientists feel that coercion, micromanagement and general overbearance of their trainees is an effective means to ensure high productivity. While research misconduct garners headlines, causing all sorts of angst upon university administrators, it might be the case that defective mentorship is actually a much weightier problem than outright cheating (Shamoo and Resnik 2003). But is it possible that these two problems could be interconnected (Anderson et al. 1997)? Mentorship is a current hot topic in science that has spawned cottage industries, self-help books and strategising among faculty members and university administrators alike. Everyone knows that finding good mentors is crucial for the young (and sometimes not-so-young) scientist wishing to be propelled into a sustainable career in the academic world of research and teaching or the private sector of research. Mentors share the unwritten rules of science. Mentors explain how these rules are intermeshed with research ethics and advise on best practices. Mentors help their students and postdoctoral trainees fulfil their dreams (should their dreams involve being a scientist). Bad mentors can shatter dreams and stagnate their trainees' careers. But perhaps even the best mentoring is not effective in deterring certain research misconduct.

Research misconduct is a major threat to science. As much as some scientists wish to point fingers at politicians and the public as the principal bad players responsible for the lack of appreciation and funding that science deserves, I think the real enemy is within our own ranks. Indeed, Brian Martin (1992) maintains that modern science, the "power structure of science," is to blame for much misrepresentation in research. Essentially scientists are not allowed to "tell it like it is" and must tell publishable stories; (he refers to the stories as "myths"). Research misconduct is insidiously damaging to the credibility of science and scientists in society since it erodes trust – not only trust in the individual researchers but in the system of science itself. Self-patrolling the profession from within is needed to reverse this damaging trend; the major pinch points for detecting research misconduct are at the levels of grant applications and manuscript review.

The ethical dilemmas in data collection, collaboration, publication and granting are likely to become even more complex and vexing in the future. More than ever, graduate students and postdocs must master more techniques, technologies and concepts in order to become and stay competitive in science. At the same time, young scientists must generate good ideas and raise increasingly scarce funds to make their research a reality. Global competition from scientists in developing countries, especially in Asia, is a new fact of life for the researchers in the West, who were formerly accustomed to the deck being stacked in their favour. At the same time, researchers in China, India, the Middle East, and other rapidly developing countries are enjoying increased levels of new funding. These new resources are coupled with even higher government and institutional expectations not only for results and publications, but groundbreaking results in publications in the most prestigious journals (e.g., Qiu 2010). From East to West, being a practicing scientist is certainly not getting any easier.

I don't wish to paint a picture of doom and gloom, however. Honestly, I can think of no more exciting time to be a scientific researcher than today with the booming innovations and opportunities to be found around every corner. We can also innovate and connect with other scientists and stakeholders across the globe in nearly instantaneous fashion these days. Certainly, the positive science news outweighs the negative news and its complications, but there is great consensus among scientists and others that the broken parts are in need of attention and fixing (Titus et al. 2008).

About four years ago, a colleague and I became convinced, for all of the above reasons (as well as others discussed later in this chapter) that a new course at my university needed to be taught on research ethics to graduate students, thus necessity spawned my new foray into ethics. After a couple of years teaching our new graduate course that met for one hour one day per week for 14 weeks, I decided that a book of this sort could be helpful to support the course (see Appendix for our syllabus), but also as a general help to young scientists just starting their research careers, and undergraduate students contemplating a career in scientific research. This book could be viewed as part guidebook, part virtual mentor, and part friendly polemic that should be helpful in addressing pragmatic problems that all research scientists experience. While virtual mentoring was part of my motivation, to substitute any book for finding a real mentor would be a mistake, which is one main reason a couple of chapters on mentorship are included. This book is on research ethics – a users' guide to success in science by following the rules that scientists largely agree are requisites for success. This book will not focus on greater issues of morality or bioethics – these are vastly different topics than the one we're embarking on here. In addition, many, if not all the chapters in this book, are subjects in their own right; the deep expertise of researchers in the social sciences, philosophy and education.

And with that, I'll state up front that I don't have all the answers. I think I do ask most of the pertinent questions, but like most things in life, asking the questions is a good bit easier than answering them. One of my main goals in asking the questions is to enable the readers to judge themselves with regards to best practices. When I started in science, I expected that there would be one right way to do experiments illuminated clearly, then analyse the data and write up the paper. It didn't take long to learn that this was not the case, and indeed, I judged myself then and ever-frequently now. Science is very creative and individualistic. There are many ways to answer scientific questions, and many ways also to go wrong. That is not to say that we can't learn from our mistakes and at least not doom ourselves in repeating the same mistakes over and over again.

So, I urge the reader to think about the questions and the answers and think about ideas expressed here, especially analysing the case studies for current and future action where applicable. Talk about these issues with your colleagues and mentors. If the topics in this book are discussed more widely in labs, hallways, and classrooms, then the best ethical practices will be advanced throughout fields of science. After I began teaching on research ethics, I found the new lively hallway discussions about various topics related to our course content was proof positive that our new effort towards promoting best practices was worthwhile.

Judge yourself

✓ Why are you interested in research ethics?
✓ What are your motivations for pursuing research?
✓ In what ways are these motivations synergistic or antagonistic with one another?

Morality vs ethics

What is the difference between morality and ethics? If morality is the foundation that ethics is built upon, research ethics is the top floor that is visible from the air. That moral foundation often has religious or spiritual ingredients and is engrained in substance that is far beyond the scope of this book. Ethics can be considered a sort of practical morality or professional morality that enables boundaries for the work of research to be played fairly. That is, if we think of problems not so much as in terms of right and wrong, but in terms of ought and ought not, then I think we understand how to parse morality vs. ethics. Many people are uncomfortable discussing morality, religion and politics. In contrast, most scientists are happy to share their opinions on ethics of their fields and science in general. It's ok if we don't all agree on the fine points of all the ethical considerations posed in this book. I worry more about the big picture.

One way to think about research ethics is in terms of best practices in conducting all aspects of research science – to maximise benefits and minimise harm. A very important ethics concept is non malfeasance – doing no harm (Barnbaum and Byron 2001). While the definitions and delineations on research ethics might seem a bit squishy, let's keep in mind that there is plenty of room for opinion. This book is about ethics much more than morality, and practical research ethics as opposed to theoretical ethics that would interest a philosopher. This book is for scientists. This book is about integrity in performing research. Summed up, this book is about scientific integrity.

Indeed, for our purposes here, this book is also about how to be a successful scientist. It can easily be argued that philosophers have thought about ethics for much longer, (e.g., Plato and other ancient Greek philosophers) than have scientists thought about science (a word not coined until the 1800s (Shamoo and Resnik 2003)). There are many viewpoints that philosophers have taken to conceptualise ethics. A few of these are utilitarianism, deontology and virtue ethics.

Utilitarianism is an example of teleological theory, which is based on outcomes rather than process. Utilitarianism seeks to do the most good for the most people; it is important to consider others and not just yourself. The utilitarian essentially does cost-benefit analysis to guide a person's path and decisions, and one that is widely implemented these days as a thought process (Barnbaum and Byron 2001).

Deontology is the ethics of duty. It strives to universalise rules that apply to everyone in guiding actions. One example here is the Golden Rule (or the rule of reciprocity), which is stated as, "Do unto others as you'd have them do unto you." "Morality as a public system" (Gert 1997 p. 24) applies to research ethics in that all scientists know the rules to be followed and is not irrational for the people who agree to participate in the system to follow the rules.

Virtue ethics focuses on living the good life. In this system, a person ought to decide to do what a virtuous person should do in all circumstances. Similar to the other two systems above, virtue ethics considers the potential for harm and avoids doing things to harm others, as this is what the virtuous person ought to do.

A last self-centred way to look at ethics is through the eyes of egoism (Comstock 2002). Egoism states that a person ought to do what is in his/her own self interests. If a scientist wants to have a long and fulfilling career, then he or she should follow the rules and perform the best science. It will also be in their own self-interest, especially in the long run, to care about others and tell the truth in science.

As a scientist, it is difficult for me to actually decide which of these various systems is most effective. To me, they all point in the same general direction to guide behaviour. If we mash them up, a virtuous scientist will seek the truth for the better good of humanity in following the rules that most scientists agree upon because it serves the self-interest of individual scientists. Scientists, by definition, should desire to maximise benefit and minimise harm (normative principles).

Inauspicious beginnings

Up until the past few years, I had no real interest in ethics as a topic of study (except a fleeting fling during my PhD training), much less in writing a book about ethics. I reasoned that everyone valued common sense ethics and there was no need to study or discuss it. When I decided to pursue science and move towards obtaining the masters, then the PhD after a stint of teaching in public schools, I was totally focused on science and research – no time for what I considered to be lollygagging in philosophical musings. In my mind, this singular focus on research was by necessity. I had found myself in so far over my head and out of my comfort zone in science, with a motivation to learn as much as I could as fast as I could. It seemed to take every drop of energy I could muster, especially in the early part of graduate training, to keep from drowning. Even then, at times, I felt I was floundering in my classes and research. I think I would have considered any training or discussion about ethics, best practices in science, or even how to *be* a scientist a real distraction from science itself. How wrong I was!

Let's imagine a fictitious mechanical engineer who is fascinated with cars. The engine design, drive train, tires, chassis, brakes, the whole thing, is an obsession. Now after studying the theory of everything automotive, our ambitious engineer designs and builds a fully functional 500 horsepower machine that's capable of going 0 to 60 mph in less than four seconds. And after all these years, our engineer will now finally drive his first car – ever – his first car being the one of his own design. Unfortunately, before taking the wheel, he never learnt the rules of the road. He doesn't know what that octagonal sign means, whether to drive on the right or left side, and let's not even consider motoring courtesies. No, our engineer considered all these things to be a distraction from what was really important – the car itself – the engineering. A disastrous crash and the destruction of the beautiful work of motoring machinery are highly likely without this key knowledge. Sad to say, the unpleasant result could have been avoided by a short course on how to drive while sharing the road with others.

While this might seem like a trivial example, it illustrates how many young scientists – myself included – approach learning science and being a scientist, seemingly by osmosis. One might argue that our automotive engineer would gradually learn the traffic laws and the accepted motoring behaviour over time, perhaps aided by a competent personalised driving instructor. But how much damage could be done in the meanwhile? As more and more students come into my lab and leave as budding scientists, I've become thoroughly convinced that learning best ethical practices earlier rather than later in a research career results in a big payout both to the scientist and the science itself. There is merit to having a driving course and a handbook.

How science works

The illustration below summarises the flow of science, at least how it is currently practiced, with all its necessary components. Science is actually a reiterative loop in which successes beget successes and failures cause the research loop to be broken. One of the primary drivers for success, as indicated by a completed and reiterative loop, or failure, as indicated by a broken loop, is scientists themselves. Having the best trained people who are eager to do research using best practices are at the heart of all successful science (Figure 1.1).

For the sake of discussion, we will designate a spot in the loop as the logical endpoint: publications. The end product of science is actually new knowledge, which must be canonised as peer-reviewed journal articles. Although there are other legitimate outlets for knowledge dissemination, such as presentations in professional meetings, books, book chapters, patents, and oral histories, the "gold standard" for credible science is peer-reviewed journal articles. This has largely been the case since 1660, when the first journal, the *Philosophical Transactions of the Royal Society*, was published.

In most cases, a science paper is built on data from well-designed experiments that test hypotheses. While professors might likely have a hand in designing

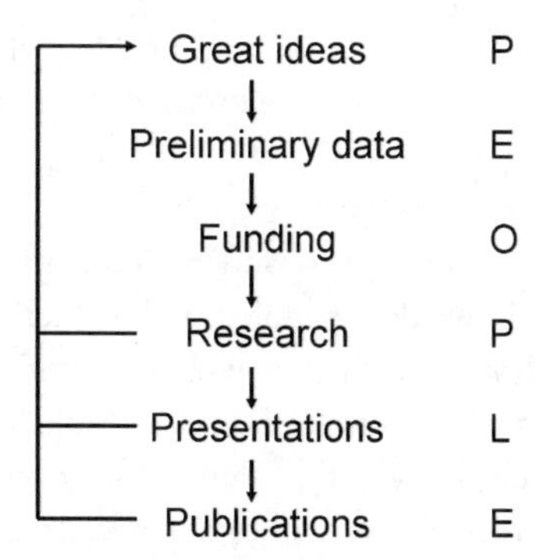

Figure 1.1 The flow of research, which starts with a great idea and background information and ends with the public distribution of new discoveries and information.

Source: C. Neal Stewart original

experiments and formulating hypotheses, it is the graduate students, postdocs and other bench scientists who actually collect and analyse data, and do most of the writing. Actually doing modern science from inception to publication is the rare luxury that few senior professors currently enjoy. While the old-professor-in-the-white-lab-coat myth continues to live in popular culture, professors are producing fewer and fewer data with their own hands in the lab; in the grand universe of data, professor-collected data are miniscule.

That's because they're busy writing grant proposals! Before I peeled away to the woods to work on this writing project today my morning was consumed with doing my part to participate in the preparation of writing parts of three separate grant proposals with three different principal investigators (PIs). The PI is defined as the scientist taking the lead role in the proposal and funded project – typically the professor or as my students fondly refer to me – the boss. None of the proposals, of course, was completed this morning. Proposals develop over weeks or months in response to requests for proposals from funding agencies. Proposal writing is so important because money is the fuel for science. In most colleges and universities, the only scientists who are typically paid from "hard" funding, that is, from university-level funding, are professors (and then again, in US medical schools, most professors are required to raise their own salaries from grants). Ironic is the world of science in that the least productive people, data-wise, are the ones who have a tenure system to protect their employment status and salary stability. Everybody else – the ones doing the work – are typically on "soft" (grant) money or term-limited funds. Why the disparity? A partial explanation is that faculty teach and are paid from university tuition income, but it is widely known that professors who attract a lot of grant funding and those with high research productivity (read, publications) are the scientists who are most esteemed in science (and by higher education administrators). In science, these professors are typically the scientists with the highest statures and salaries. Again, why? They are the ones who enable the funding of science to collect the data to publish the papers. Famous papers containing groundbreaking science in turn yield status to institutions (and more money), thus the financial circle is completed. Universities successful in research have greater reputation and funds enabling them to get even larger coffers, hire more faculty members and continue the trend of the rich getting richer.

If money is the fuel of science, ideas and preliminary data are the drill and refinery, respectively. Without ideas coupled with sufficient data to demonstrate that the ideas are sound and worth pursuing, it is difficult, if not impossible, to find appreciable funding for science research. Funders are generally a risk-averse lot. It is a long-dead myth that famous scientists can get funding on the basis of their name-recognition alone. Science does not allow the resting on laurels. To remain successful, scientists must continually generate good ideas for grant proposals. Do they do that alone? In most cases, no. They get help from postdocs and graduate students to make science get started and go-round: ideas → funding → data → publications (Figure 1.1).

One can see two potential problems arising already. First, many critical steps are being performed by young scientists-in-training who might be inexperienced with both the ethics and politics of science, not to mention the nuances of the science itself. Hence they could simultaneously be targets for exploitation and temptation. Tales abound of graduate students who are taken advantage of and not treated in such a way that their professional success is enabled. Second, each of these steps toward publication can be stumbling blocks where scientific and ethical problems might arise. Therefore, addressing potential ethical dilemmas in the context of modern scientific practices should be of some practical help to students just beginning this journey. In fact, I argue here that understanding the rules of science are necessary for running a laboratory and research projects. Subsequent chapters will build upon these themes.

Summary

Arguably, science is the most exciting and invigorating of pursuits and careers with seemingly endless opportunities to create knowledge. With the increased funding and emphasis of scientific research occurring worldwide, science is also growing increasingly complex with opportunities for funding and publications becoming more and more competitive. We can think of research ethics as a framework for creating a fabric of integrity, which should, in turn, make research findings stronger and the researcher happier.

Judge yourself *redux*

✓ Why are you interested in research ethics?
✓ What are your motivations for pursuing research?
✓ What ways are these motivations synergistic or antagonistic with one another?

In the redux sections I will offer some of my own reflections. For many of these "judge yourself" questions there are no right or wrong answers, but the questions are designed to help the reader ponder his or her opinions and feelings about pertinent topics.

Most scientists with whom I've spoken about research ethics are not as interested in ethics as an academic discipline as they are in the result of ethics, which is to say, robust and trustworthy research results. The best scientists I know are driven by the quest for knowledge and they are eager to grasp an understanding about how things work. There are many legitimate reasons to become a scientist, including curiosity, autonomy, and the opportunity to research topics of interest. It is important to measure motivations as you mature. With more at stake, career motivations can change, and sometimes not for the better. However, often motivations become more noble and less self-centred. It is important to continually judge your own motivations with regards to actions – to see oneself clearly as if looking into a clean mirror.

Chapter 2

How Corrupt is Science?

ABOUT THIS CHAPTER

- More than ever, science is in the public eye.
- Science is funded mainly by public sources and therefore held publicly accountable.
- Many scientists admit to dubious and unethical behaviour.
- Older scientists misbehave more often than younger scientists.
- Unethical behaviour can reach beyond activities classified as research misconduct, which is defined generally as fabrication, falsification and plagiarism (FFP).
- FFP carries strong sanctions.

This chapter title itself seems corrupt, bombastic and inflammatory. Substitute a number of other professions for "science" and few people bat an eye. How corrupt is . . . auto repair, or entertainment, or law, or politics? Justified or not, any of these seem to, at least, sound better than insinuating that science can somehow be a crooked pursuit. Like the arts, science seems to be one of those unassailable undertakings in which the pursuer has a higher calling; where idealism and truth trumps money and comfort. Ask any scientist, "Are you in it for the money?" While a few scientists have found the field to be quite lucrative, invariably, no scientist I've ever posed this question has answered in the affirmative. Contrast this with many other fields requiring a great deal of education. True, scientists typically don't enter science for the money, but motivations can change, and with these, behaviours can also change between the beginning and end of a career. Besides money, there is also the factor of sheer survival in a field populated with creative and smart people. Corruption can also be borne from the notion of gaining an unfair advantage and the "publish or perish" culture (Woolf 1997).

In addition to examining motivations, we need to assess how widespread is scientific misconduct. Does breadth necessarily define impact on true discovery and science itself? That is the key question. After all, perhaps science corruption is akin to cheating on taxes. Little money is at stake for a low wage earner, but if a billionaire cheats, then significant funds are in play. Of course, there is a huge

Research Ethics for Scientists: A Companion for Students, First Edition. C. Neal Stewart Jr.

practical difference between plagiarism on an undergraduate assignment and fabricating data sets on a *Nature* publication that is highly visible. But on the other hand, cheating is cheating, and little-to-none should be tolerated in science when found-out.

Judge yourself

✓ Why did I decide to enter science as a profession? What were my motivations?
✓ Am I generally tempted to cheat? In what ways?

"Scientists behaving badly"

A number of surveys and meta-analysis studies have been conducted on research misconduct through the years. For example, Ashworth and Bannister (1997), Eastwood et al. (1996) Fanelli (2009), Hard et al. (2006), Maurer et al. (2006), Pryor et al. (2007), Swazey et al. (1993), and Yank and Barnes (2003) have all examined behaviours of scientific misconduct from data taken by surveys. I wish to focus much of the following discussion on just one survey for simplicity's sake, yet I invite readers to dig deeper into the literature to judge for themselves the breadth and depth of the problem. In 2002, a survey was mailed to US National Institutes of Health (NIH) grant recipients, which asked them to anonymously self-report research misconduct by yes/no responses to several questions (Martinson et al. 2005; see also Anderson et al. 2007). The survey was sent to mid-career scientists who had received their first full-sized (R01) grants – these were typically associate professors, having a mean age of 44 years. It was also sent to younger scientists who were supported by NIH postdoc fellowships. On average junior responders were 35-years-old. From a few thousand surveys mailed, 52% responded from the older group and 43% from the younger group. To this point, I am struck by two surprises already. First, a high proportion of scientists were willing to report their misdeeds. Even if respondents could maintain anonymity, why risk self-reporting bad behaviour? Confession is good for the soul, true, but why take the chance of being discovered for data fraud or other bad behaviours? Second, the sampling of scientists chosen as the study subjects was highly skewed. NIH (R01) grant recipients are nearly always very well-qualified scientists who are serious about performing biomedical research. The scientists surveyed were no dilettantes. R01 grants are simply not easy to win as the current funding rates approach just above 10% (http://report.nih.gov/nihdatabook/); and the postdocs receiving NIH fellowships are no slouches either. These postdocs represent former graduate students whose grades, aptitudes and research are among the best in US biomedical fields. Prior to reading the results, and based upon these initial conditions and assumptions, I would have been shocked to learn that either group reported much bad behaviour; I just would not expect them to participate in research misconduct or in suspect research practices. What did the survey report?

Before we look at these particular results, the prevailing opinion, as reported by Martinson et al. (2005), was that incidence of falsification, fabrication and plagiarism in science is probably in the 1–2% range. But in 2004 the editorial office of the *Journal of Cell Biology* (Rosser and Yamada 2004) estimated that papers containing questionable data might be as high as 20% (Anonymous 2006). One editor of a Chinese "campus" journal recently reported that 31% of submissions contained plagiarism (Zhang 2010). While longitudinal data on cheating doesn't exist, most people in science would agree that if we do have a 20% incidence of misconduct, or even "questionable data," then we have a huge research integrity problem; then there is plagiarism (Butler 2010).

According to Martinson et al. (2005) three researchers out of every thousand admitted to fabricating or "cooking" data. Cooking refers to altering existing data to "improve" a finding rather than outright data fabrication. Others' ideas were used (plagiarism) without proper credit for 1.4% of the respondents. For these two items, there were no differences between mid-career and early-career scientists. There were differences for at least two notable questions, with mid-career scientists being worse. Twenty-four of a thousand older, as opposed to just eight of a thousand younger, scientists used supposedly confidential information in their research. One would hope that as scientists age, they would adopt better practices and not worse. A big offence, 20.6% mid-career scientists admitted to caving in to pressure from funding sources to alter their experimental design, methods, or *results* (emphasis added) as opposed to 9.5% of early-career scientists. Herein we have a striking dichotomy. Most scientists would agree that while they might be passionate about their science, scientists ought to approach research dispassionately and objectively since the main objective of science is the pursuit of truth. However, there is a public-held dogma that funders with profit-motives, or that ideological motivations of funders often drive or alter research results that are reported. The high numbers of researchers altering experimental design, methods, and especially results, demonstrate this skeptical viewpoint of motivation has validity. I might add that the NIH is among the most benign of funders in this regard; their grantees should feel no ideological or economic pressure to find one result over the other. It would be interesting to perform the same survey among recipients of funds from pharmaceutical, agricultural and chemical companies, where a research agenda is more obviously skewed toward economic impact.

If researchers are found to be guilty of many of the items above, there would be university and government sanctions levied against the perpetrators. But we actually hear of very few cases in which guilt is discovered and penalties are exacted. Most misbehaviour in science seems to go undetected. One of the reasons why this is the case will be seen in Chapters 6 and 7. Scientists are not fond of non-anonymously reporting blatant misconduct or even sloppy science; it seems that few people want to be known as whistleblowers, or in the childhood vernacular, tattletales. This is understandable, and in the above survey, 12.5% of scientists admitted to overlooking others' flawed data or their own "iffy" interpretation of data. This figure doesn't even take into account any close examination of papers by peer-reviewers looking specifically for misconduct. Therefore, we see that many scientists would rather look the other way than "objectively" report on research honesty – even if they could do so anonymously, which is often the case in peer reviewing of grant proposals and publications. While scientists might grumble over bad players during the social hour, they are not anxious to call out their peers, even if not doing so results in damage to science as a whole (Gunsalus 1998). It would seem that the proverbial rug covers a profundity of scientific dirt.

The Martinson study illustrates that the magnitude of incidents and self-reporting scientists with dubious behaviour can hardly be considered insignificant. In fact, results in some of the categories are startling. What is even more profound is that the survey asked scientists to report on their behaviours during only the *past three years*. Furthermore, the sheer frequency of misconduct is staggering. Martinson et al. (2005) reported that best estimates of falsification, fabrication and plagiarism prevalence were thought to be in the neighbourhood of 1–2% – indeed this is in the range of a meta-analysis of FF (not P) surveys (Fanelli 2009). In the Martinson survey, however, one-third of respondents admitted to deeds that research officers would consider to be sanctionable! For early career scientists, the frequency was 28% and for mid-career scientists it was 38%! Therefore, it is almost unimaginable to estimate how high the proportion of scientists is who are guilty of gross unethical behaviour over a scientific lifetime. What if they had surveyed even older scientists? If we extrapolate, well over half of researchers nearing retirement would participate in bad behaviour during their final three years in science. Consider the fact that the average mid-career scientist in this survey had perhaps 20 or more years left in his/her career, the opportunity and propensity to participate in dubious activities is overwhelming. The survey selected relatively young scientists who had, seemingly, spent little time in science. And let's remember, that the numbers reported here are likely conservative since it is doubtful that all the guilty parties would self-report their misdeeds for fear of some sort of retribution.

Do scientists behave worse with experience?

Anecdotally, I've observed that many scientists become more savvy of the rules of science to the point that they know which ones they can break without being caught. The Martinson survey indicates this might be systematic among biomedical researchers. Of the 16 questions posed, 6 of these had statistically significantly different responses among mid-career and early-career scientists. In each case, the older scientists reported higher incidence of misconduct compared with younger scientists. The conclusion we must draw is that age and experience are important factors causing scientists to go bad. As Lutz Brietling opined in *Nature* later in 2005, younger scientists more likely have higher ideals and enthusiasm compared with older scientists who might become jaded. "In the rough world of today's science, they are exposed to an environment in which impact factors and awards are more important than advancing the knowledge of mankind." (Brietling 2005). Sad to say, these same scientists are also probably running roughshod over their trainees. He thinks that the practice of science itself, as defined by how we do modern science with its pressures and rewards, is the problem; at its root, disillusionment is the problem. While I somewhat agree with his conclusion, I think the ultimate root of the problem must be deeper still than simply mere disillusionment. Breitling further states that he doesn't think sanctions are the answer to the problem. I'm not so sure I agree with him. In fact, another letter writer to *Nature*, Kai Wang, thinks that education along with stiffer penalties would go a long way toward improving scientific integrity. Wang, a graduate student, points out that there is little ethics education in graduate school (Wang 2005).

Judge yourself

✓ What factors of science research might cause you to cheat? For example, how do you deal with pressure?
✓ How might these be counteracted?
✓ How much of a deterrent is embarrassment and punishment?
✓ How do the results of the Martinson survey make you feel about the profession? About yourself?

Crime and punishment

Important to addressing research misconduct is reporting and sanctions. Nobody, it seems, is anxious to police science. Editors, to some degree, are the most proactive players in science in this regard, but clearly, peers, and especially students, are not anxious to make waves. As we'll see in Chapter 7, whistleblowing often comes at a steep price. But as Wang (2005) succinctly points out, "If the benefits of misbehaving outweigh the possibility of being punished, academic misbehaviour is probably inevitable."

Is scientific misconduct inevitable? Unfortunately, to some degree I think it is indeed unavoidable inasmuch that corruption is present in every profession; why would science be immune? That said, we shouldn't abandon high standards of expectation for honesty and refuse to stem the tide of unethical behaviour. What can be done? First, as has been mentioned as at least a partial solution, education of the expectations and rules of science is crucial (Titus et al., 2008; Titus and Bosch 2010). Second, we must be aware of factors leading to potential disillusionment and corruption in mid-to-late-career. Third, scientists need to self-police the profession more effectively. I am not referring to the state of mind to which Nobelist Marie Curie abhors: "There are sadistic scientists who hurry to hunt down errors instead of establishing the truth." Quality assurance is critical in verifying that published data and information are honest and real; I think that's one duty of science and scientists. One subdiscipline in science publishing that is emerging is informatics tools to catch cheating – either pre- or post-publication. Algorithms and routines to spot plagiarism and illustration manipulation exist and should improve. Journals should be the vanguards of these activities since they arguably have the most to lose by publishing papers containing FFP or dubious results (Berlin 2009; Butler 2010).

This leads us to a discussion of consequences for being caught violating common professionally-accepted standards. In contrast to many commentators, I don't think, for most violations, that the penalties should be "fatal" or completely debilitating to the scientist. Banning someone from accepting federal funds for 3–5 years might be an appropriate penalty for many deeds of misconduct (McCook 2009). I think that we probably underestimate and underutilise the power of professional embarrassment, or at least, the threat of embarrassment, which could be a very effective deterrent to scientific misconduct (Errami and Garner 2008; Berlin 2009; Koocher and Keith-Spiegel 2010). Clearly, the scientific community needs to debate reporting, investigating, and penalising those who participate in research misconduct with the hope and expectation of rehabilitation.

One vehicle for scientific embarrassment is when a paper is retracted, especially from a high-profile journal. Often papers are retracted by the authors when an honest mistake is made; these are understandable and not sanctionable. Nath et al. (2006) estimate that just over 25% of retractions are the result of research misconduct. Though it is not the majority of cases, it is an important fraction. In a landmark case, *Science* retracted, without the authors' permission, a paper that contained fabricated data (Normile 2009). While it took the journal four years to decide to do this, and after they tried and failed to contact all authors to obtain their permission for retraction, it still took gumption to take the action of pulling the paper. Journals might ask authors to retract papers, but, until now, journals have waited until the authors made the decision themselves to pull a paper. This is a good policy. It is easy to imagine a disgruntled scientist who takes retribution on his co-authors by asking a journal to unilaterally retract a paper. Journals just don't do that – papers, in a sense, are transcendent over their authors and their authors' wishes. In this particular case, some of the authors of a 2005 paper agreed that the paper on tracking proteins should be retracted because data were fabricated. The method to track proteins that was presented, called "MAGIC," is still valid and should be patentable according to the senior author of the paper, even though the published results and data were indeed fabricated. Muddying the water even further in this case is that a company was started by the same author and others to commercialise the method. The waters here appear very muddy indeed. However, one thing is certain, the authors, especially the senior author, are paying dearly for their misdeeds. The institution where the researchers were employed, the Korean Advanced Institute of Science and Technology (KAIST) in Daejeon, South Korea, has taken the dramatic action of dismissing lead author Tae Kook Kim. Furthermore, his company has sued him for fraud. KAIST is now developing educational programmes to ward off trouble and to "promote research ethics and integrity." (Normile 2009).

Should misconduct always be punished so harshly? A case can be made that leniency is often appropriate and that harsh sanctions are unfair. In the US, the Office of Research Integrity (ORI) is the investigative and sanctioning body for research misconduct by NIH grantees. The charges they investigate fall under the heading of fabrication, falsification and plagiarism (FFP). If found guilty, researchers end up listed on their "Administrative Actions" website and the guilty are banned from obtaining federal funding (debarred), typically for three to five years, but according to Allison McCook (2009), once the information hits the internet, it never disappears. She spoke with several formerly listed investigators who were found guilty of relatively minor research infractions. These researchers claimed that their misdeeds have resulted in their being forever stigmatised. One example was an assistant professor at a large public university who included data in a grant proposal that he really didn't have authority to use. His story was that he prepared a grant proposal using preliminary data from a collaborator at a startup company – one that he'd recently worked with on another project – who was to be a co-investigator on the proposal in question. A few days before the proposal was to be submitted the collaborator informed him that the company didn't wish to participate. At that point the assistant professor should have removed the data in question but forgot to do so, even though he did remove the collaborator from the proposal. The company found out that their data were included in the proposal and complained to the assistant professor and then the university, which was then required to forward their findings to the ORI. The assistant professor admitted that he was guilty of research misconduct as charged. Although his story ends relatively happily (he received tenure a year after he signed papers admitting guilt and is a productive scientist), other researchers struggle with the stigma of similar-magnitude misconduct. McCook (2009) chronicles researchers whose stories ended with prolonged ramifications – not so much by the ORI sanctions, but because the sanctions are forever searchable on the internet. Like naughty pictures of sophomoric high jinks posted on Facebook that come back to haunt job applicants and beauty queens alike, research indiscretions have long-lasting consequences. One researcher has had a difficult time finding research employment despite a rich publication record. Out of desperation he changed his publication name from his real name to escape the stigma. Another researcher left academic science, experienced stress-related health problems and then later started a small lab-based company in his kitchen.

Judge yourself

- ✓ How do you feel about "crime and punishment"?
- ✓ What can be done to create a more accountable science environment in your lab?

So, on second thoughts, I'm not convinced that perpetual embarrassment is the solution to research misconduct, especially for relatively minor offenses. I think that education on best practices can be the most helpful intervention for young researchers (Titus and Bosch 2010). Slowing down to think about ramifications of research and its representations are also useful. Certainly the punishment should fit the crime and experience and history should be considered. Do we live in such a corrupt science culture that we can't trust most research outcomes? I think the answer is "no." In spite of reports of high-levels of misconduct, I believe that most published science is sound. None of the McCook cases mentioned above involved retracted papers. Often a discerning eye finds data that "doesn't look right" or results that are too good to be true, and it does not hold the credibility as better papers. Between discernment and replication of experiments, science truly is a self-correcting enterprise, at least most of the time.

This chapter focused on making bad ethical decisions and not incompetence. Honest scientists make all kinds of judgments about which experiments to perform, what data to collect and report, and what data should remain essentially invisible to the public. These decisions are not relevant to misconduct but rather are components of professional judgment. Non-scientist Napoleon Bonaparte teaches us well, "Never ascribe to malice that which can adequately be explained by incompetence." There are as many levels of competence and shades of grey as there are numbers of scientists. Anyone who works in biology, for example, knows that there is great variation in biological systems and that some outliers don't represent the reality of the research accomplished. Sometimes, simply bad science (incompetence) and fatal errors are the cause of retraction of publications instead of misconduct. Indeed, when I review papers, I typically look for honest mistakes in experimental design, execution of experiments and data interpretation. That is really the purpose of peer-review. Misconduct, as defined by FFP, requires the intent to deceive to be "proven" (Shamoo and Resnik 2003). Indeed intent to deceive is often hard to prove.

Joseph Macrina (2005) discusses the question, "Who needs ethics?" (Weston 2002) in his book and frames the answer in light of the science profession's needs themselves. It is obvious to me that the needs are greater than we could've imagined. The readers of this book will delve into the rules of science and learn how ethics guides the brightest path while building a successful research career. We won't worry so much about falling into the hole of FFP, but how ethics can help us stay on high and solid ground. After all, most of the research ethics decisions that scientists make are not so much, What data will I fabricate today? Or, Which paper would be the best one to plagiarise? But rather scientists must most often judge their own actions to check for objectivity and pure honesty in doing science.

Judge yourself *redux*

✓ Why did I decide to enter science as a profession? What were my motivations?
✓ Am I generally tempted to cheat? In what ways?

Most scientists are curious, creative, and want to do experiments to learn more about the world around them. They never get tired of learning. Of course, money is needed to live, and sometimes, with the world of science being competitive, there are temptations to cheat in any number of ways – from cutting corners to outright fraud.

Judge yourself *redux*

- ✓ What factors of science research might cause you to cheat? For example, how do you deal with pressure?
- ✓ How might these be counteracted?
- ✓ How much of a deterrent is embarrassment and punishment?
- ✓ How do the results of the Martinson survey make you feel about the profession? About yourself?

A big issue currently is pressure to publish in international journals, obtain funding to survive in science (keep a job). Science can be stressful and it is important to deal with stress in productive ways. It is also important to stay current with knowledge in our respective fields. I would personally be greatly affected by embarrassment – maybe even more than punishment – because it is important that I feel respected in my field. I feel like I have tried to build a good reputation and would not want to see it tarnished. I don't want to lose good will amongst my peers. The Martinson survey makes me examine myself in the various categories to see what my weak spot might be. Then I want to bolster that part so I don't fall prey to misconduct. I really do want to have an untarnished record in science.

Judge yourself *redux*

- ✓ How do you feel about "crime and punishment"?
- ✓ What can be done to create a more accountable science environment in your lab?

I think I understand both crime and punishment. I have a streak of justice in me that goes along with a streak of fairness. I want to think that in the end, justice is served. I think that it is important to talk about all these issues of integrity and to encourage labmates to have integral lives, which include integral professional lives. I also think that it is important to be "in touch" with my lab group and be careful about travelling or staying sequestered in my office with the door closed too much.

Summary

Research misconduct is a significant problem in research science with about an incidence of 33% self-reported unethical behaviour within a three-year period. Everyone in science, including scientists, reviewers, editors and publishers are responsible for ensuring accurate scientific information is reported and also responsible for holding high standards for integrity. Sanctions for misbehaviour, especially FFP, include disbarment from funds and public and professional embarrassment.

Chapter 3

Plagiarise and Perish

ABOUT THIS CHAPTER

- Plagiarism, defined as using others' ideas, sentences or phrases without citation, is perhaps the most common form of research misconduct.
- Plagiarism is easily avoided.
- Plagiarism is becoming more easily detected thanks to computational and networking tools.
- Some self-plagiarism is typically unacceptable in instances where the source material has been published.
- Recycling your own writing is usually acceptable when it has been previously unpublished.

Plagiarism is claiming others' ideas, sentences, or phrases as one's own. In professional writing, plagiarism is fraudulent and intentional by the FFP definition. In student writing, plagiarism might be done without any malintent. For example, students with a lack of command of language and science also could be tempted to plagiarise as a survival mechanism. Nonetheless, in science, fairness and honesty dictate that others' work is recognised and cited in scientific literature, grant proposals, and coursework. This is the widely accepted standard among all academics; scientists or otherwise. Plagiarism cannot be tolerated or condoned.

To me, plagiarism is the most boring of ethical offences; largely borne of both academic laziness and ignorance, and being contented to remain in this state. Or, in haiku:

> Cannot synthesise?
>
> Lazy about your writing?
>
> Why not plagiarise?

Shamoo and Resnick (2003, p. 50) refer to authorship as "perhaps the most important reward in research." But then follow to declare that "publish or perish. . .is

Research Ethics for Scientists: A Companion for Students, First Edition. C. Neal Stewart Jr.

a grim reality of academic life." In our youth the teacher assigned a maybe dull theme paper topic. However, we choose our own research and have the privilege to share our hard-earned results with the world through publications. Thus, I agree with Shamoo's and Resnick's first statement about writing being rewarding, but I admit that I can't grasp how enthusiastic scientists could ever view publishing as "grim." Therefore, my view of plagiarism is the case of having to fake the acts of creativity and communication. Indeed, communicating results is vital to scientific research (Macrina 2005). Of course, plagiarism ranges from mild-to-extreme, and I think that the extreme cases are typically the only ones uncovered. Mild plagiarism is seen, I think, oftentimes as a legitimate shortcut that is often tolerated – and there are various reasons for this. First, no one really wants to police science writing, and so only blatant plagiarism is reported and punished. Second, some suspected plagiarism is vague. Third, a measure of self-plagiarism is acceptable and even useful in science. No matter what your view on plagiarism, there are electronic tools that are being increasingly used to discover it in professional and student writing, and so reporting and punishment will become more important in the future. These developments and predicaments will be examined in more detail later.

There does seem to be various cultural views about plagiarism (Pecorari 2003; Scollon 1995), but it is clear that in science, there exist international standards for "owning" one's writing in which plagiarism is not allowed. Mimicry in nature often confers and evolutionary advantage, but in scientific writing it can be a career killer. One special problem worth mentioning is cryptomnesia. In the context of plagiarism, a cryptomnesiatic writer might believe an original sentence is his or hers, but in reality it might have been learned earlier (Brown and Murphy; Pecorari 2003; Pennycock 1996). Thus, it is conceivable that some plagiarism is inadvertent. This might happen more with students who learn by memorisation. In addition, in some cultures imitation in writing, i.e., plagiarism, might be construed and accepted as a form of flattery, but in research and coursework at universities, it can lead to dismissal. A professional but otherwise plagiarising scientist can be subject to censure by employers, journals, professional societies and granting agencies. This chapter will examine what plagiarism is and what it is not, and it will examine grey areas in scientific literature (e.g., using the author's own papers as source material) where there seems to be a fine line between right and wrong. Plagiarism abounds. Errami and Garner (2008) estimate that perhaps 200,000 of the 17 million papers indexed in Medline, a database of life science papers, might be duplicates. Some of the 200,000 are outright plagiarism, but perhaps 30 times more are simply duplicated articles by the same author. Is this an allowable practice?

This chapter will seek to define plagiarism in order to, hopefully, avoid it altogether, and definitely point us toward best practices in professional science writing. As you might imagine, plagiarism is not a new topic. It has been examined from many angles (e.g., see Carroll and Appleston 2001; Carroll 2007; Harris 2001; Howard 2001; LaFollette 1992; Maurer et al. 2006; Pak 2003; Scollon 1995).

Ideas

Idea generation is crucial in science. Lacking sufficient original ideas is a problem that might be insurmountable when attempting to sustain a career in research. Good science is defined as the implementation of novel ideas and sound methods to address pertinent questions and problems. Indeed, these, to a large degree, define what is fundable in grants and contracts. Without money, there will be little science. But ideas and lines of scientific research are continually recycled aren't they? To some extent this is true, but there should always be an increment of novel contribution to warrant publication in the scientific literature (for the most part anyway, as we'll see later). Perhaps more than an increment is needed to win grants. Certainly, if an idea has been proffered by someone else and you mention it in a paper, the source should be cited. This gives credit where credit is due and also impresses your peers that you know the literature. No one likes someone who pretends to have thought of everything first. Science is about ideas and data building on existing ideas and data – Newton's notion of standing on great shoulders coming before him.

Sentences

Copying and pasting sentences from one paper to another is plagiarism. The surest way to get caught is to also include internal referencing from the source paper. For example, Paper A contains the following sentence:

> What distinguishes GFP from other reporter genes is its ability to fluoresce without added substrate, enzyme, or cofactor (Prasher 1992).

Now if we see this particular sentence verbatim in Paper B, citing Prasher 1995 and without citing Paper A, then plagiarism has occurred. This is an open and shut case. Even if citing Paper A, the author of Paper B would need to put the above sentence in quotation marks, and, in practice, this is almost never done. The author of Paper B would be attempting to convince us that he knew this fact and was clever enough to write this sentence. People often get caught plagiarising if their own writing is visibly weak and an author includes plagiarised sentences of a vastly different style and quality.

Phrases

Copying or the apparent copying of phrases is not as clear cut as the copying of sentences. That is because certain phrases are optimal expressions of ideas or descriptions, and so we're not surprised to see common phrases reoccurring in the literature. People get suspicious about plagiarism if familiar phrases get used quite frequently in a paper, especially if they are from the same apparent source.

If possible, specialised phraseology should be cited, especially if this denotes a unique contribution from someone else. This is a grey zone, which calls for judgment about specialised knowledge and phrases.

A hoppy example

In the same issue of the regional specialist journal *Southeastern Naturalist*, there appeared two unrelated papers on swamp rabbits. One paper was out of Arkansas and the other from Illinois.

Fowler and Kissell (2007) begin the introduction of their paper saying, "*Sylvilagus aquaticus* Bachman (swamp rabbit) is primarily found in bottomland hardwood forests along rivers, streams and swamps (Allen 1985, Chapman et al. 1982, Kjolhaug and Woolf 1988)."

In the second paper in the same issue Watland et al. (2007) start their paper with, "*Sylvilagus aquaticus* Bachman (swamp rabbit) is a representative species of bottomland hardwood forests, occurring primarily in swamps, river bottoms and lowland areas (Chapman and Feldhamer 1981)."

For each of these papers, I examined the cited references to answer two questions. First, was there any plagiarism? And second, were the references appropriate to cite? To answer the first question, there was no plagiarism. Each topic sentence by Fowler and Kissell (2007) and Watland et al. (2007) were distinct from those in their references and just happened to converge on similar wording by chance. This example teaches us that happenstance is a kinder and more accurate explanation than plagiarism in many instances. To answer the second question, several general papers written in the 1980s about swamp rabbits describing their habitat and habits were cited. Each one I surveyed had seemingly appropriate features to make them citable in this context. It could be argued that the fact of swamp rabbits living in the swamp could be obvious and general knowledge, but the authors of both papers played it safe by citing appropriate references.

What is plagiarism, really?

So, aside from these textbook definitions, what is plagiarism, really? When do people get concerned that a paper contains plagiarism? How much originality is needed to define a work as original, and how much similarity can be tolerated before it is declared plagiarism? Let's start with ideas. Very little in science is totally novel and plagiarism of ideas is the hardest to prove. By its nature, most science is derivative and incremental, but it is important to cite appropriate literature from which a paper might be derived.

Judge yourself

- ✓ What is your ethical background with regards to plagiarism?
- ✓ Were you taught about plagiarism in school?
- ✓ How good are your communication skills and command of the languages that are commonly used in science, especially English?
- ✓ Are you comfortable and confident in writing?
- ✓ Do you think you have good ideas?
- ✓ What is the best way to improve your science and writing?

Indeed, it is rather easy to find apparent plagiarism if one wishes to look hard enough. Most scientists have things they'd rather do. But, if we spot a phrase or sentence that appears to be familiar, one of the easiest ways is to enter it into Google or some other search engine within quotation marks and analyse the results. Let's look at an example, namely the definition of plagiarism at the top of this chapter. The definition I offered was one that was burned into my memory from my Chief Justice days in the early 1990s (see preface for the story). So, I did a Google search of: "Plagiarism is claiming others' ideas, sentences, or phrases as one's own." I was genuinely surprised to find that a Google search of this sentence in quotations yielded no results. So I tried smaller phrases until the transposed "ideas, phrases, or sentences" finally turned up one hit. Additional transposing of the words returned other hits. So, is my original sentence plagiarism? No, because I used my own knowledge and memory of the subject. Alternatively, I could have simply looked up a definition and quoted and cited it. Maybe that would have been safer (but I would not have generated this example). Also in my defence, I had no intent to deceive. At its core, plagiarism is a misconduct characterised by misrepresentation. Also in my defence too, one could argue that the general definition of plagiarism falls under the category of common knowledge, and so it would be difficult to prove plagiarism in such a case. For example, it is not plagiarism to write, "The sky is blue." That is the case, even though this sentence has certainly been written in hundreds or thousands of works. In science there are any number of grey cases that could fall either way, but the general expectation is that the author will provide new knowledge somewhere in a paper; in best cases, a lot of new knowledge. Fields of science are composed of knowledge that ranges from "something every good physicist knows," for example, to obscure and arcane knowledge that should always be cited.

How many consecutive identical and uncited words constitute plagiarism?

This is the question many people ask and I don't think there is a clear answer, although other people obviously disagree with me – different people have different rules. The University of Idaho Information Literacy module on plagiarism states that the use of others' two or more words in a row should be placed within quotation marks (http://www.webs.uidaho.edu/info_literacy/modules/module6/6_4.htm). I find "that criterion" to be untenable, "albeit conservative" to catch all "possible plagiarism." This system would result in distracting quotation marks littered throughout the document. So, I think two is too few. I've seen other anti-plagiarism enthusiasts make the case for three-, four-, five-, and more words in a row to be problematic. I tend to think finding the bottom end of exact phraseology to be the wrong focus. The right focus, in my opinion, is to focus on good and original writing. When authors want to use the perfect sentence or phrase that was coined by another writer, then they should put it in quotes to be safe. In most cases, authors should simply put phrases and sentences in their own words and provide an appropriate citation.

Self-plagiarism and recycling

Plagiarism is using others ideas, phrases and sentences without proper attribution or citation. **Not allowable.**

Self-plagiarism is plagiarising your own published work; typically significant amounts of copied and pasted segments. **Mostly not allowable.**

Recycling is copying and pasting your own writing from an unpublished source to another document that might be published or unpublished. **Generally allowable.**

Self-plagiarism is simply using your own words in more than one document; typically the source document has been published before, hence the plagiarism (negative) label. There is a wide variety of views about self-plagiarism, ranging from "duplicative publication is unethical because it is wasteful and deceitful" (Shamoo and Resnik 2003, p. 83) to the notion of the impossibility to steal (plagiarise) from yourself. The safest haven in writing and the best practice is to equate self-plagiarism as equivalent to plagiarism (Green 2005; Berlin 2009). The duplication of material among papers from a notable scientist in Canada has prompted investigations and the call for new rules by the Natural Sciences and Engineering Research Council (Reich 2010). In this case conference proceedings articles were double published later in peer-reviewed papers, which led to the retraction of some of the journal articles. There is certainly a functional difference between recycling your own writing when it has been unpublished previously and self-plagiarising from a published paper work that will be used in another published paper (double publication). Recycling is allowable and is not self-plagiarism. The functional scalpel here divides between published and unpublished work, where duplicate publishing is unethical.

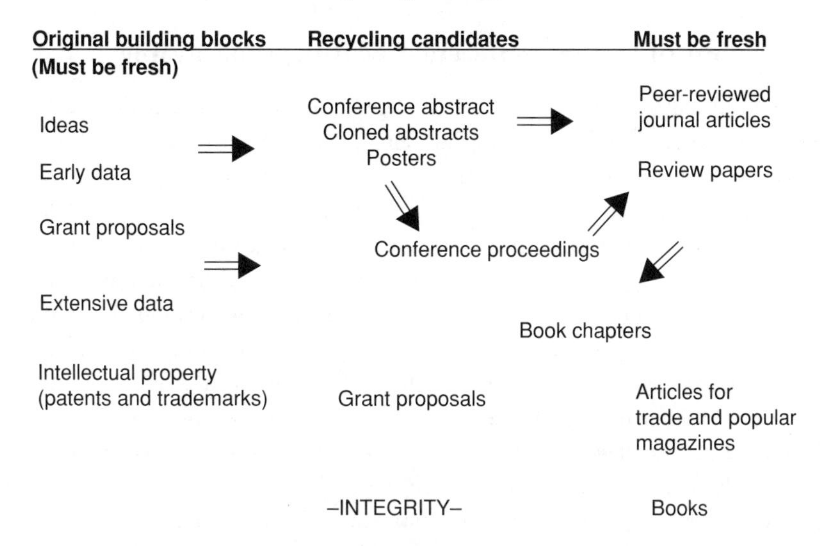

Figure 3.1 The writing recycling network. This proposed network demonstrates some likely and ethical sources for recycling your own writing as denoted by arrows for source-sink relationships. The final products of research in the right hand column should seldom be used for recycling. Even if they are used as sources, text should be largely rewritten, updated and improved to avoid self-plagiarism.

Source: C. Neal Stewart original

I think that a common sense approach about sources and sinks is helpful. For recyclable sources, such as your own grant proposals, notes to yourself, and unpublished white papers: the key criterion is that they are not published. But as sources move from non-published to published, the verbatim copying should be avoided. For sink documents, i.e., the document that work is being pasted into, there is more leeway for allowance for recycling without ethical ramifications. The goal is for each published document to be unique and containing new and valuable information. Whole manuscript duplication should be avoided, unless it is simply a reprinted manuscript and indicated as such in the front matter. Copying and pasting among certain documents of your own is completely allowable as long as the source is unpublished (Figure 3.1). It is also best practice to cite any source published documents that were used during recycling in the sink document; e.g., the peer reviewed paper (Reich 2010).

Judge yourself

✓ How do you feel about copying and pasting?
✓ How do you feel when copying and pasting?
✓ What is your view about self-plagiarism?

Path of science publications

Recycling of much conference writing is allowable

After an idea is hatched and data are collected, the consummate publication, the one that "counts" in the eyes of other scientists and the people evaluating your career is the peer-reviewed publication. Do your best writing here. But many (most?) times a peer-reviewed publication is born as an abstract for a talk or poster at a scientific meeting. For a talk, the abstract is the only writing that might represent a project to the public. A poster might contain the draft writing that will morph into a scientific paper that will be submitted for publication. The same is true for a thesis or dissertation. Therefore, your own abstracts, poster text, and dissertations can be completely recycled into publication material. That is, feel free to pull text from any of your conference abstracts and posters to help write your own peer-reviewed publications. An efficient scientist presents a poster that can simply be flipped into a publication with a little tweaking. If that resulting publication is a peer-reviewed publication, then the self-plagiarism must come to a halt for that particular writing stream. The peer-reviewed publication is the dam. Conference proceedings articles are often considered as writing "halfway houses," also fair play for self-plagiarism. At least that's the case in many fields. Why are these allowable sources? Conferences and conference proceedings are considered to be appropriate vetting for ideas and results that is perhaps not quite mature or complete. In the life sciences, conference abstracts and proceedings are generally not considered to be the finished products, but rather places for projects to receive their first rounds of criticism so that they can be refined for the peer-reviewed publication that comes later; this is really helpful to authors. Ever since the days that typewriters were shelved for word processors no one writes everything from scratch, and it is important to determine what documents are worthy to build upon and which ones should be torn down and buried. You must make wise judgments for yourself. And you are only allowed to build upon your own abstract or poster; not someone else's. In certain fields, such as computer science and engineering, many conference proceedings are peer-reviewed and constitute the final publication in the stream. This almost never happens in my own field of biology. Therefore, best practices vary among fields and you have to know the particular rules and best practices in your particular field. It is therefore not surprising that the software package, SPLaT (Collburg et al. 2003), for the detection of self-plagiarism, conceived as a tool to help prevent self-plagiarism in computer science. The reasoning behind SPLaT and the argument against self-plagiarism is nicely stated on the SPLaT homepage (http://splat.cs.arizona.edu/).

> "It is our belief that self-plagiarism is detrimental to scientific progress and bad for our academic community. Flooding conferences and journals with near-identical papers makes searching for information relevant to a particular topic harder than it has to be. It also rewards those authors who are able to break down their results into overlapping *least-publishable-units* over those who publish each result only once. Finally, whenever a

self-plagiarized paper is allowed to be published, another, more deserving paper, is not."

Patents

After scientists gain more experience, they know a good publication or good publication nugget when they see one; a paper that leads to a noticeably substantial contribution to science. A publication idea might not be given first at a conference. In fact, a really hot paper might be presented at a conference only after the publication has been submitted because of patent issues. For patentable inventions, it is best practice not to publicly disclose information – "prior art" in patent lingo – before a patent is filed. A publication could very well enable someone else to reduce to practice your invention. Of course, this is the whole point of scientific publications: to disseminate knowledge so that it can be beneficially used by others. Filing for patent before the knowledge is disseminated protects an inventor's commercial rights by protecting patentability. It is perfectly acceptable also for a patent application to have self-plagiarised material – even though a patent attorney is writing the patent application on the inventor's behalf.

Review papers and book chapters

Again, as a scientist develops a reputation as an expert in the field, there will arise numerous invitations to author book chapters and review papers. These are often good opportunities to contribute to the foundations of science. Review papers are also peer reviewed and these count towards a scientist's dossier of peer-reviewed publications; sometimes in its own category. Sometimes the same is true for book chapters, but these are not created equally. Certain books, it seems, are published by obscure publishers and don't seem to receive wide readership or citations. Other books seem to be taken much more seriously as citable source material. For our purposes here, we won't judge the prestige of the chapter or the book and how much it will be read and cited. Rather, we will attempt to develop best practice guidelines about self-plagiarism and writing recycling guidelines for these documents as sources or sinks.

Both review papers and book chapters are opportunities for an author to synthesise old information into a new publication in which existing knowledge in the literature is coalesced into summaries and opinions about a field of research and where it might be headed. A truly well-written review paper has value in that it synthesises new insights from previously published papers. Bad review papers read like shopping lists of who did what, but without the synthesis. Reading review papers are certainly a recommended best practice for young scientists trying to learn the broad landscape about a subject: the platform to launch new science within a context of what's already known. Since reviews are peer-reviewed (but not all book chapters), they are also entered into the canon of the scientific literature. Inevitably, after a good review paper is written, someone editing a scientific research book will ask for that same information to then be reworked for inclusion in said book. Certainly it is acceptable to do this, as long as self-plagiarism is avoided. For example, whenever I'm asked to author or co-author a book chapter, it is nice to start from one of my prior review papers or a grant proposal as a skeleton document about what is known in the field at the time. The expectation from most book editors is that a skeleton document is a good place to start, but nobody wants to see a reproduction from source to sink; self-plagiarism. It is not acceptable to copy and paste the latest review paper, change the title, rework the referencing format and call it a new contribution. Typically my colleagues and I will do extensive editing to rewrite sentences, change the flow of an existing paper by altering the emphasis and headings, and update the references so that the new contribution has value on its own merit. Many of the ideas may have some resemblance of those found in the prior work. That's fine, since the ideas included in the review paper are typically the ones a book editor wants included in the book in the first place. But care should be taken not to simply reproduce another work. Furthermore, there should be no intent to deceive the reader. It is important to cite the review (source) to alert the reader of the prior published paper in the later paper. See the self-plagiarism/recycling network below.

Grant proposals

In several instances in this chapter, grant proposals have been mentioned. Writing grant proposals to support the funding of research is a necessity in modern science. A government agency will issue a call for proposals, and scientists respond with their ideas and plans about how to address the call with their proposals. While it is not okay to plagiarise another author in your proposal, recycling your own writing is perfectly acceptable when writing grant proposals. I've recycled any number of writings into grant proposals, and that is useful to express the idea of the proposed research in a time-economic fashion. Grant proposals are not publications. In fact, they are considered to be privileged documents for review purposes only (see Chapter 9). New writing certainly goes into grant proposals, but that can be legitimately copied and pasted into posters and publications alike or other grant proposals written by you. Once the peer-reviewed publication is written, it can be recycled into additional grant proposals, but not into new publications. The idea behind grant proposals is to not disseminate new knowledge, but to propose an idea and project for funding. Therefore, your own recycling of text describing methods sections, literature reviews, rationales and justifications is often useful when writing proposals. The ethical dilemma that we'll see later is that it has to be for work that is non-overlapping. That is, one could envisage using the same methods and techniques to address a new problem and project. Grant proposals are special cases in the world of writing. Even though proposals are not publications, per se, it is not acceptable to fabricate data or plagiarise others in proposals. Nor is it acceptable to plagiarise ideas. These would all be unethical and sanctionable.

Writing for coursework

Since grant proposals are not real publications and are available as sources or sinks for self-plagiarism, it make sense that papers written for high school, college, and university coursework could fall under the same category and be treated under the same rules as grant proposals. If that is the case, why do professors get upset when students plagiarise or self-plagiarise? The reason is, even though themes and term papers are not real publications, the expectation from all professors is that students will produce something new and related to specific elements in the course. The big idea is that performing literature searches and reviews, synthesis, and the process of writing new material for assignments comprise a substantial act of learning. Therefore it is not allowable for students to self-plagiarise their own writing for courses – no learning or real writing takes place when that occurs. Cheating in this way defeats the whole purpose of learning to write and synthesise ideas for student development. It is not a shock that some colleges and universities permanently dismiss students who are found to be guilty of plagiarism.

Self-plagiarism and recycling summary

To dissect the differences why plagiarism and self-plagiarism are unethical and why certain text recycling is ethical, we need to revisit the previous chapter on the foundations of ethics. One benchmark of ethical decision-making is based on minimising harm and maximising benefit. For plagiarism, the person whose ideas, sentences, and phrases are stolen has been violated and the thief takes the credit for the presented scholarship. Both people are harmed. To test whether individual cases of copying and pasting is self-plagiarism (unethical) or recycling (ethical) we need to investigate potential harm. If the reader expects a totally new synthesis, such as the case of a review article in a journal, then no self-plagiarism is acceptable. Since edited books are often collections of previous ideas that are compiled into a single volume, there is more lenience for recycling, but not self-plagiarism. If a scientist were to publish multiple research papers that are self-plagiarised and contain identical or near-identical data (salami slicing – see Box on pp. 34–35), then science and society is harmed, since it is assumed by the scientific community that each paper should not be repetitive. Each case can be analysed in this way to identify sources and magnitude of harm to judge what the best actions should be. With little practice, most scientists easily discern best practices in their fields.

Judge yourself

✓ How would you feel about someone plagiarising your work or stealing your ideas?
✓ How much do you want to trust authors and their papers?
✓ How would you feel if you discovered an author had published duplicate papers on the same material in two different peer-reviewed journals?

Dissection of a manuscript for plagiarism

A scientific manuscript is typically composed of five sections: 1. abstract, 2. introduction, 3. materials and methods, 4. results, and 5. discussion. Tables, figures, and references are also included as well as a short acknowledgments section. The abstract is typically the last part of a paper to be written, so it should always be freshly synthesised to fit the content and tone. The introduction section might contain ideas and progressions that are common to many papers on similar subjects. For example, I used a derivation of a sentence published in 1996 in a multi-authored 1997 paper. Both of these papers were about applications using the green fluorescent protein (GFP) in transgenic plants. GFP had only been used in a few studies in plants, and only beginning in 1995.

My 1996 paper: "What distinguishes GFP from other reporter genes is its ability to fluoresce without added substrate, enzyme, or cofactor (Prasher 1992)."

My 1997 paper: "GFP is the only well-characterized example of a protein that displays strong, visible fluorescence without any additional substrates or co-factors (Heim and Tsien 1996)."

Note that the idea is the same in both sentences, but that the second sentence is constructed better and says more. Note also that the reference was updated to reflect the new scientific development. In both papers, the other novel properties were discussed in the same order. This is not plagiarism. Even if the sentence from the 1996 paper was not mine, the 1997 paper would not have been a case of plagiarism.

Like the introduction section of a paper, the descriptions found in the materials and methods might also be similar to other papers by the same or even different papers. Chemical names, formulae, instrumentation, organisms, geographic locations, etc. might be quite commonly-used among closely related papers or papers using similar methods that authors should describe. One might argue that citing a paper in which these items were perhaps first listed could be more appropriate than a verbatim listing. If it is a common scientific procedure that has a memorable label and is at least somewhat routine, then this practice in entirely appropriate. But since the purpose in science is to communicate clearly about what was done and learned, then oftentimes, lists and descriptions are necessary to include as a service to the reader. Again, think of the Golden Rule. What practices do you want used in papers that you read? What practices increase clarity? Indeed, there is often only a few logical ways to phrase certain things and these might be repeated from time to time. This is typically not construed as plagiarism. But don't copy and paste methods sections from one of your (or others') papers. Certainly the safest route is to cite the best paper(s) for a method and then offer any updates or modifications briefly in the new paper.

Results and discussion sections should always be written fresh and special care should be taken not to plagiarise. The discussion section is always the most difficult to write since the author is putting the new results in the frame of reference of existing published knowledge. Typically, if there are sentences that seem familiar in a discussion, then it warrants investigation of possible plagiarism. Since it is such a tough task to write a good discussion, this is a logical place for the lazy plagiarist to plagiarise. This is a shame because the discussion section is the best opportunity for real creative expression in a scientific paper.

So, as we see, similarities among ideas, sentences, and phrases must be placed in context of each paper and the sections of each paper. Similarities are not created equal.

Tools to discover plagiarism

There are a number of anti-plagiarism services and tools on the market. The following companies host websites that describe for-fee services:

www.turnitin.com

www.ithenticate.com

www.plagiarismscanner.com

www.plagiarism-detector.com

www.canexus.com

www.plagiarismdetection.org

One can also use Google or other search engines to find text that is questionable. There is no doubt that publicly-available scripts will be written to help catch plagiarism.

Publishers are increasingly using CrossCheck, which is available to join by members of CrossRef (http://www.crossref.org). Anyone who has read scientific papers online has noticed CrossRef in the reference sections. CrossRef is a cooperative among publishers that allows the cross-referencing of citations

(Butler 2010). CrossRef yields a network of references and citations that helps link scientific literature across publishing companies. CrossCheck screens candidate text against full-text of published papers to alert publishers against possible plagiarism. Note that this would also find inappropriate self-plagiarism. It also provides similarity indices between the reference paper and published papers to help editors find reviewers and also as an aid for determining the uniqueness of a submitted paper. It is a helpful tool to determine if an author has cited all the closely-related papers to the current study. As of this writing, publishers were checking approximately 8000 papers per month using this tool (Butler 2010).

Self-plagiarism and ethics revisited

The accepted definition and repudiation of plagiarism is clearer than that of self-plagiarism, although there is strong agreement that self-plagiarism should be avoided. Judicious recycling for unpublished sources is allowable. For example, if I am asked to make four presentations in one year to different audiences about the same topic or study, I am probably going to recycle parts of the abstract and subsequently update and refine it. To make a totally new abstract each time is a waste of energy and not useful to science or society. There is agreement that self-plagiarising papers and proceedings is very detrimental to science (Green 2005, Berlin 2009). However, allowances should be made for refinements of published proceedings articles to morph into peer-reviewed journal articles. Self-plagiarism should be avoided and science and society benefit. Ethical considerations of harm and benefit should be minded as we perform writing assignments.

Judge yourself

✓ How do you feel about your papers being scanned computationally for plagiarism?
✓ How would you feel if plagiarism was detected and your university administrators were alerted?
✓ Do you have a trusted colleague you can ask to read drafts of your papers and make an "off-the-record" judgment about your writing and instruct you on plagiarism?

Salami slicing

Salami slicing is the practice of double publishing or publishing several papers from a study instead of fewer larger and more comprehensive papers. Salami slicing, also known as the "least publishable unit" could lead to repeating (plagiarising) methods, data, and introductory materials, and unnecessarily dividing up datasets; all of which are unethical (or if not unethical, certainly not best practice). Salami slicing is not in the best interest of science as the result is convolution and confusion of scientific results. If a unified dataset is divided into salami papers, it defies logic and expectations among the community of scientists. The best practice is to publish papers in "natural" units, whether they be small or large papers. I've often published techniques papers that are very short. I've also published long papers with lots of results that fit together in one unified story. Sometimes it is better to imbed a new technique within the context of a bigger paper and sometimes not. It is important to consider the ethics and readability of parsing up a study rather than keeping it intact if it would better serve science to keep all the subtopics together as a unit. What would the reader rather read – several smaller papers (or a subset of these) or one bigger and more comprehensive paper? Which is more useful? One growing problem in science is the "avalanche of low quality research" (Bauerline et al. 2010). This avalanche refers to papers that are written but never cited – more than half of scientific papers are cited within their first five years of publication. Let's let ethics be our guide – why publish a paper that no one reads or cites?

Is plagiarism getting worse?

I present no data here, but I think the consensus answer among researchers is generally "yes," with plagiarism being greatly enabled by full-length online articles and the internet. Plagiarism is to academics as steroids are to sports. Cheating becomes easier because of technology, and there seems to be an escalation in prevalence and perhaps tacit acceptance. I've heard plagiarism defended as acceptable because of the difficulty of non-English speakers having to communicate their science results in English, but this is not an acceptable excuse. As in most technology-related crimes, detection is always playing catch up, but it is heartening that more and more journals are using tools like CrossCheck so they can avoid publishing plagiarised material.

The case of the plagiarising graduate student

A graduate student (Michael) was in an ecology course. In the sizable class, Michael sat toward the front of the class on the rare occasion he attended; these were the only seats available by the time he arrived. He was a snappy dresser and easily noticed. Michael was one of the few PhD students around with much money. While he imagined himself as a scholar, he enjoyed golf more than his studies.

This fictionalised case occurred just before the introduction of the World Wide Web, when word processing was done on mainframe computer terminals and the results were printed off remotely in the computer science building. Therefore, typewriters (and typists) were still alive and well, and the industrious student who was endowed with fast fingers could earn some extra cash as a typist. Typically, the trade was advertised on bulletin boards with rates and phone numbers listed on tear-offs at the bottom of the page.

Michael happened to, unbeknownst to him, choose such a typist who just happened to sit in the middle of his political science class: Sally. In addition to being a fast and accurate typist, Sally was also a very good political science student and an observer of people. Michael, being neither observant nor often present in class, did not recognise Sally. Therefore, when Michael met Sally for the first time, it was to hand her a composite of published pages that were cut into sections and taped together, with the instruction to type this "for a friend." While Sally might have inquired about the unusual nature of the request and its propriety, being the consummate professional, she did not. She merely received the pages and did her service. She did, however, recognise her new typing job as the assignment that had been given three weeks prior in their class.

In a week, when the assignment was completed, Sally made a photocopy of her typed work-for-hire, and then gave Michael the original, which he promptly turned in as his term paper. As one might imagine, by the week's end, Sally had grown angrier and angrier at Michael; the result of having typed her own term paper that she wrote herself as well as Michael's, which was, of course, plagiarised. He was clueless, of course, until he was notified that he was being investigated for plagiarism. Michael was incensed that Sally had spilled the beans, and he offered no real defence during his honour court hearing. He received an "F" for the course and probation. Within the year, however, he had withdrawn from the university and moved into a non-academic pursuit.

This story is illustrative inasmuch that Sally would not be contacted today to type Michael's project, and therefore there would be no one to report his bad behaviour. The reader would have had to pick-up on disparate writing styles to detect. The students of Michael's ilk of the world know this and plagiarise all the more. I'm certain of it. All they have to do is cut and paste, just as Michael did, but using the computer, and there is nobody who knows they are doing it.

Judge yourself *redux*

✓ What is your ethical background with regards to plagiarism?
✓ Were you taught about plagiarism in school?
✓ How good are your communication skills and command of the languages that are commonly used in science, especially English?
✓ Are you comfortable and confident in writing?

✓ Do you think you have good ideas?
✓ What is the best way to improve your science and writing?

I was taught very early on about the evils of plagiarism, cheating, etc. That does not mean that I never did these things, but I was never caught. My language skills have, until recently, been poor; even though English is my first language. My writing is not bad now – good enough for technical writing but not good enough to be a novelist. Through massive amounts of writing, I have become a better communicator. When I was young I did not plagiarise because of poor English skills. I plagiarised because I was lazy, even though I always have had more ideas than I could follow-up on.

Judge yourself *redux*

✓ How do you feel about copying and pasting?
✓ How do you feel when copying and pasting?
✓ What is your view about self-plagiarism?

I love copy and pasting. In one way it limits the number of errors that can be created and it is fast. If the source I'm copying from is allowable I feel great, but as it moves more into the grey zone, my conscience tells me that I should abandon the strategy and just write from scratch. My view of self-plagiarism has changed over the past few years from it being ok to it not being ok – all for the reasons described in this chapter.

Judge yourself *redux*

✓ How do you feel about your papers being scanned computationally for plagiarism?
✓ How would you feel if plagiarism was detected and your university administrators were alerted?
✓ Do you have a trusted colleague you can ask to read drafts of your papers who will make an "off-the-record" judgement about your writing and instruct you on plagiarism?

I don't enjoy being plagiarised. I have experienced it and it makes me feel angry. At this point, if my writing was found to have been plagiarised or if I was caught plagiarising, I would feel that due process should take place and responsibility accepted. I would feel sad either way. I used to use readers when I was unsure of my English, but I never asked anyone about scanning for plagiarism. A number of computer programmes can do this, but young scientists could probably benefit from reading each others' papers.

Summary

Plagiarism is probably the most frequent form of academic misconduct. Both plagiarism and self-plagiarism is unethical and should not be practiced. If some sort of systematic inadvertent plagiarism, e.g., cryptomnesia, occurs with regularity, its root cause should be addressed. Unlike plagiarism and self-plagiarism, text recycling from your previously unpublished writing is acceptable in science publications. Computational tools are being increasingly used and will most likely catch serial plagiarisers.

Chapter 4

Finding the Perfect Mentor

ABOUT THIS CHAPTER

- The choice of a mentor is a crucial decision for the young scientist.
- Deciding on a mentor is more important than the choice of institution.
- There are several signs of good and bad mentoring that can inform the decisions of students and postdocs.

The perfect mentor doesn't exist, of course, but a good mentor can help launch a young scientists' career. A bad one can kill or cripple it, and encourage mediocrity. This chapter is written mainly to help the young scientist: selection and then, "training," of a mentor, and to provide guidance and tips when someone subsequently becomes a mentor, which is bound to happen when scientists run their own labs. But actually, the point in time when that someone actually becomes a mentor is open to interpretation. I'd like to think that graduate students and even undergraduates can aid the mentoring process in helping more junior members of the lab gain useful experience and savvy. I've observed this in action with great results. The ethics of mentorship are complex. On the one hand, laboratory directors (mentors) must be selfish in one sense in that opportunities must be prioritised to maximise results and build their own and the lab's reputations. On the other hand, lab directors have an ethical obligation to nurture graduate students, lab staff, and postdocs – the people who actually do the experiments, write the papers and even participate in grant proposals. Good mentors make their trainees' dreams come true, and don't become their students' worst nightmare. Good mentors find no dichotomy in performing the best research where trainees are nurtured and mentored so that someday they too, can fledge into their own labs.

My view of mentorship is likened to one-on-one teaching and enabling junior mentees to also mentor. Mentorship also includes the relay of professional ethics to the young scientist and especially how to become an independent researcher, which includes a plethora of "secret" rules and nuances. There are many ways a mentor can go wrong, and the student has to be proactive in choosing and

Research Ethics for Scientists: A Companion for Students, First Edition. C. Neal Stewart Jr.

keeping a mentor active in good mentorship. The mentor is the hinge that swings the door for young scientists-in-training. But the mentee often has to push the door open. As doors open or close, it becomes clear that choosing a mentor is one of the most important decisions to be made in science, even though often it is greatly underappreciated and overlooked.

Caveat

Ok, I don't think that I'm a super mentor: with empathy faster than a speeding bullet and insight able to leap tall personnel problems with a single bound. But, good mentors abound, and I am certainly grateful for mine – those scientists who supervised me during graduate and postdoc training, but also those people afterwards who helped me learn the systems inherent to science and universities. There seems to be no shortage of peculiar politics in all institutions. Many successful scientists tell stories of having successful mentors. Much of the information in this chapter arises from my own mentoring mistakes and not those from my own mentors. My material also comes from anecdotal stories and observations from others. Names and places have been changed.

The first step a graduate student or postdoc faces is choosing the right mentor. It also is increasingly common for assistant professors to be assigned a faculty mentor to help them become accustomed to their institution and department, as well as giving them advice on tenure and promotion, so this chapter is applicable to a wide range of scientists in various stages of their careers.

Choosing a mentor

Aside from the chosen field of science and a potential project, the mentor is probably the most important factor that shapes the initial direction of a scientist's career. The trick is to find the mentor working in a setting and doing research who is compatible with and compelling to the student – and to find a professor with funding! Science without funding is like a car without petrol – not moving. Graduate students in science are nearly always paid a stipend to help them obtain their advanced degrees, but money is not always universally available. Finding the best major professor suited for a particular candidate has been greatly aided by the internet, and fortunately, it has never been easier to locate and correspond with candidate professors. On the other hand, faculty members are inundated with emails, sometimes not personally addressed, requesting opportunities for study. A universal process for finding graduate school opportunities does not exist, but generalised guidelines for finding an advisor are straightforward. Specific advice where students and postdocs are most mobile is more applicable to the US.

Home away from home: the university, department and program

Making the optimal choice for all three of these items above is important – just not nearly as important as choosing a mentor. A student will be linked forever to the university where the PhD was earned. Whenever you give a guest lecture, seminar, or any other presentation, you will hear the name of the university as part of your introduction. Your PhD hood that you'll wear for university commencements is identified with the university and college. Sometimes the PhD robe is also unique (e.g., the crimson robe of Harvard University is always easy to spot in graduation exercises). Your reputation and the reputation of the university and department/program are indelibly linked, making these important considerations. But I argue here that the importance of finding a good mentor outweighs the reputation of programmes or universities, because the mentor, unlike an institution's reputation or facilities, directly shapes the student in myriad ways.

Some students want to attend a particular school for location, reputation, amenities, and other reasons (e.g., "my dream is to have a degree from 'Exclusive U'"). In many parts of the world, there are limited choices for students constrained to study in their own country. While these justifications are not invalid, they add unnecessary constraints that can muddy the waters of deciding which lab to join. Not to fear: the best mentor will seldom be located at the worst university or worst department. I think the best search scheme to find major professor candidates is to determine who are the top 5–10 scientists in your particular research area of interest, and then match the entrance requirements of his/her associated department and program with your own qualifications. After all, the major professor's lab is where the student spends the most time, and it is the major professor who will be giving out advice and shaping the research project and subsequent publications. The quality of the science or engineering performed coupled with mentoring will be the major determinants of the next step in a career, along with the success enjoyed. In many programmes, the major professor also has a strong voice on which students get admitted via funding and advocacy.

The major professor

Start early. Choose wisely. Be persistent. In the best case, the choice of major professor can propel a career to the stratosphere. The student will be the first author on multiple major research papers published in excellent journals, and will learn, first-hand, what a successful career in science looks like. In the worst case, the wrong choice can drive a student out of science, or worse. One achievement that pleases me most is that the overwhelming majority of my ~75 trainees are still in science and enjoying fulfilling careers.

The first step toward becoming a scientist is knowing the field of interest by reading pertinent journal articles. Computer-based literature searches (Medline, etc.)

of keywords to find journal articles that are interesting are a good first start. Once you find articles of interest, determine which person in the author list is the professor driving the research. Usually papers have several authors, but few professors. Typically, but not always, the principal investigator (the "PI") is the "corresponding author" of the paper; it is important to study with people with good ideas. After a list of prospective major professors is made, the next step is to find each person on their university websites, typically listed by departments. Then read all the web pages and papers authored by the person. Short of stalking the major professor, you must learn all you can about this person. Students should respect and be interested in the mentor's research, and then it helps to actually like the person too.

A big mistake is to enroll in graduate school or a take a postdoc position without actually visiting the prospective mentor and the facilities. There is no substitution for setting foot in labs and offices, shaking hands and looking eyeball to eyeball. During a one-on-one meeting with the professor, it is important to gage how this person will perform as *your* mentor. For example, how does this professor deal with interruptions? Is the prospective mentor warm or cold?

It is imperative to learn how students are funded, and for how long. Beware of the PI using the phrase "my grant" or "my grant proposal" as in the singular. A lab that operates on a single grant is typically not stable funding-wise since the grant might or might not be renewed every three years – the research might not continue. Where does this leave the student? All the better if the student can get his or her own funding.

There are many things to discuss. Certainly, the PI's philosophy and practice of authorship should be discussed during a visit. And most importantly, the student should meet and talk with other graduate students in the lab to determine how pleasant and productive your experience will be if you join. Facilities and equipment are important as are materials and supplies. If budgets are always tight, then students and postdocs will be frustrated. During an interview, you really want to determine if you can envision yourself working in that setting with the people you meet. Lab colleagues are nearly as important as the major professor – in certain cases, more important. You'll work with many of them shoulder-to-shoulder each day while the boss is sequestered away in an office.

> "When I arrived to his office, he was fiddling around with his computer and was obviously frustrated with some software glitch. He begged for me to be patient while he fixed this one problem prior to our scheduled meeting. His frustration level and temper rose until he finally abandoned his computer issues and turned to me. His tone suddenly changed as he went from frown to eerie smile. The smile reminded me of a character from a vampire movie. With that we began to talk about research, which I found quite interesting: both in theory and methodology. We were having a good conversation and I was imagining myself working with him. Then,

> his graduate student showed up at his door and simply stood there. She seemed to be expected but she did not interrupt our conversation. Finally, he noticed her and, his demeanour changed again. He asked quite tersely for her to pick up four specific books from the library that he needed for a grant proposal. "Make sure they are on my desk before lunchtime" he said. After the books arrived and our meeting concluded, he took me out to lunch and promised me a good stipend and the project of my choice. It should have hit me square in the face, but it was only on my way home that I realized that something did not seem right about joining his lab. I guess I would not want to be treated the way he treated his current student. And I wouldn't want to spend the next few years with a Jekyll and Hyde type of boss, or worse, a vampire who'd want to suck me dry. Getting a PhD can be something of a horror—who needs the unnecessary drama?"

It pays to heed the vibe! It is smart also to determine whether a PI's current students and postdocs are happy in their research or whether, given the chance, if they would choose another mentor. Other questions are worth asking. Do students graduate with their intended degrees? And how long does it take? Are people happy? Where do people go when they graduate? Where do postdocs typically go once they leave the lab? Life in science is, surprisingly to many people on the outside, predominantly about people and their interactions, namely the mentor-student and lab member-lab member relationships, which shape the environment and success of the mentors' labs and their trainees. Ethics and best practices can be affected by the lab vibe.

> "The potential for the breakdown of trust underscore the need to look more closely at relationships in research and laboratory groups, and at the ways in which groups are managed. A research environment in which relationships are distant, frayed or fractured – an unhappy lab – may well not sustain responsible research conduct. Trust is of key importance to the enterprise of science" (Weil and Arzbaecher 1997).

Judge yourself

- ✓ How good a judge of character and environments are you? Can you "size-up" people? How rapidly?
- ✓ Are you an "easy sell" or able to resist promises that are too good to be true?
- ✓ What are your most important features in a mentor and laboratory? How does your personality and style play into your choices?
- ✓ How do you feel about mentor accessibility? Do you prefer to walk in and chat when someone's door is open or would you rather make an appointment? Are you a texter or an emailer? What is the mentor's communication mode and style?
- ✓ Are you a trusting person? Do people tend to find you trustworthy in return?

The two big things that mentorship should impact: funding and papers

The most important scientific metrics are associated with productivity, namely the BIG TWO: grants and papers. If a mentor can encourage a trainee to become independent in these two items, and to approach them with respect and integrity, then good mentorship is occurring and there is a good chance the trainee will have a fulfilling training experience and career.

Authorship

It is crucial that there is a clear understanding about a mentor's style and practice about funding and authorship. In a perfect world, graduate students are first authors on any papers stemming from their graduate projects and will be middle authors on side projects in which the students have made significant contributions. Alternatively, some major professors prefer to take complete control of papers and thereby deny students opportunity for independence and growth. There are quite a few PIs somewhere in the middle of this continuum. Related to this, prospective students and postdocs should determine if professors prefer to publish lots of shorter papers with few authors or fewer grand papers with many authors, or again, something in between. One scenario is that several graduate student projects might be needed to make a grand paper that might be mainly controlled by a postdoc or the PI who would then likely be the first author. Never lose sight that the students and postdocs should protect their interests in publications as authors. The bottom line is that for young scientists to develop in science and progress to the next career stage successfully, the number and quality of papers authored and authorship position are very important factors. Papers should be a reflection of the student's expertise, research productivity, and real contributions.

Funding

For many graduate students and postdocs, research funding "just appears" and is never a worry. In some unfortunate cases, funding is not stable wherein a student is not aware of the magnitude of instability until the money is gone and a student is "self-funded," i.e., not funded at all. Maybe graduate students and postdocs should play an active role in obtaining funding and are partly responsible for reporting research progress to agencies or sponsors; but it is good to make sure there is a high probability of stable funding over the degree program. If young scientists are to be successful in research, then they must become familiar with the world of grants and contracts. What better place to start tuning-in than as a PhD student, and then continue to fine tune grantsmanship during a postdoc stint? Many mentors require trainees to produce portions of proposals that get rolled into a larger document; mini-proposals of sorts. In other instances, young scientists can land their own small (or not-so-small) grants and become masters of their own universe. Landing my first (and second and third!) small grants while I was a graduate student was as joyous to me then as the big grants I get now. At the time, $600 from Sigma Xi and $1000 from a company was crucial to paying for supplies and a computer for my PhD project. At the time, I was paid from a nine-month teaching assistantship and was working on a project that was otherwise not funded with a federal grant. It also gave me confidence in playing the grant game. I learned I liked it and was pretty good at it. In fact, success in grant-getting is quite often tied to winning tenure in many colleges and universities where research is valued.

It is important to keep in mind that, ideally, the project you adopt as your own should be fundable, if not already funded. Science is fuelled mostly by government grants. In the USA, the NSF and NIH fund a miniscule portion of those proposals submitted each year. Therefore, successful scientists must necessarily be proficient at getting their grant proposals funded. It is fair game to ascertain how proficient your potential major professor is in the grant game. Ask for a list of awards. Shy away from someone who won't give you a list of grants or talk funding. I've seen grant ruses played on occasion and the student is the one who gets pinched. I know of a faculty member who has claimed he's had multiple grants from a certain agency and a search of the agency's database fails to turn up any grants where he was the lead PI. Students must practice due diligence to parse the fakers and dilettantes from the real deal. Doing fundable research is your ticket to having a chance at being a successful and independent research scientist. Students burdened with unfundable research typically will not be afforded the opportunity to reach their potential. Some of the best advice I ever received as a graduate student came from a professor on my graduate committee. He frankly informed me that my masters research, while probably publishable and interesting to me, was definitely not fundable nor of wide interest to many other people. He also offered that people would pay me to be interested in several other topics if I stretched my own interests to pursue something more fundable for my PhD, and then even more fundable (funded even!) for my postdoc. A teaching assistantship allowed me to follow my nose in seeking research topics and my mentor provided a lot of latitude in research. This stretching me out of my comfort zone also led to my gaining new skills and expertise. Research is typically a winding path, and not a highway. Successful scientists are able to master new techniques and develop their own thinking about ideas that are the best to pursue. The successful graduate student is adaptable and grows up to be an adaptable senior scientist. When beginning a graduate degree, it also helps if adaptation happens rather quickly.

> "One of my best friends in graduate school was so much smarter than me. In biochem, she was the first one to finish homework sets (and her answers were always right). Her grades were stellar. I felt really bad for her when they kicked her out of the graduate program after her second semester. They gave her the boot because she would never decide on a PhD project. Picking a project, selecting a committee, writing a research proposal, and giving a talk about the proposal were all things that had to be completed before the end of the second semester. The big stumbling block was deciding on a single project. She was just indecisive I guess."

Final thoughts for graduate students and postdocs in mentor selection

There are a few additional items to consider when selecting a mentor. It is a measured risk to select a young assistant professor as a mentor. The risk is that the mentor does not get tenure and is fired from the university. Typically the system demands that the assistant professor is either "up or out"; i.e., getting tenure and a promotion to associate professor or out of a job. In the case of the latter, the graduate student typically becomes a sort of ward of the university and must scramble to finish a degree, and the postdoc is simply out of a job. So, it is wise that trainees have strong cause to believe that young mentors will be awarded tenure. If there is any hesitancy from the assistant professor about his or her chances in getting tenure and promotion to associate professor, then buyer beware! The same is true for other systems that use contracts – the student benefits if a professor's contract is extended and not terminated. Sometimes faculty members just move to another location. If this happens, there is often good opportunity to remain with the mentor at a new location. When I moved to my present position, nearly my entire lab elected to move with me and some of these people had significantly better pay and project support. None fared worse. Finally, on the other end of the spectrum, you'll also want to keep an eye out for age, health, and retirement prospects of a potential major professor. Your career will not be accelerated if your mentor retires, dies, or spends lots of time in the hospital during your graduate program.

There is also the consideration of mentor success and lab size. Malmgren et al. (2010) performed an analysis on mathematics mentors and the fate of their trainees. They found that trainees of mentors with lots of trainees had more trainees themselves. They also showed a correlation between number of trainees and productivity. Another interesting result was that there was a benefit to being a trainee of the young successful professor compared with the older successful professor. Therefore, it is beneficial for a student to pick a successful mentor, but there is a greater benefit to picking the mentor early in the mentor's career. I'm not sure how a person can know very early in a career how much a mentor will succeed later, but this study confirms what many people have suspected all along in mentorship: success breeds success.

Choosing a graduate project

Certainly, the consideration of what project to do for masters and PhD research is as important as choosing a mentor. Sometimes in choosing a mentor, the choice of project is already made. That is, the mentor might not have any choices for a new student than one specific project prescribed by a grant. In most cases there is some flexibility. As Reis (1999) quotes department head George Springer, "It is really important to do the right research as well as to do the research right. You need to do "wow" research, research that is compelling, not just interesting." I think this is really good advice in that it inspires students to aim high and not sell themselves short of their potential. Of course, this book is more about doing the research right than doing the right research. What project is the right project depends on many factors, especially the individual student. At the end of the chapter there is a case study that works through many of the relevant factors.

Judge yourself

✓ How aggressive or assertive are you? Do you feel like you take control of situations?
✓ Are you a good judge of character?
✓ Do you have trusted friends and colleagues whose opinions you trust?
✓ Do you listen to advice?

Mentors for assistant professors

In many departments, new assistant professors are assigned established faculty mentors. In many ways, mentorship to assistant professors is far less crucial, career-wise, than for postdocs and graduate students. Why? Since obtaining an assistant professor position is probably the biggest employment bottleneck in all of science, anyone who has got that far has probably already received valuable mentoring. Therefore, chances are that assistant professors have enjoyed a certain amount of success and independence already during their training. That is sure to carry over into the tenure track career as a faculty member and earning tenure, which is generally related to having success in grants and publications.

A sage mentor or group of mentors can help an assistant professor land that first big grant and become even more productive and independent. Alternatively a bad mentor or a toxic department can inadvertently drive the mentee to another university or out of science. I've seen both good and bad scenarios in action. It is paradoxical and unfortunate that some departments acquire the reputation of "eating their young" by bad mentoring. No one wants to be eaten. This figuratively occurs when mentor(s) attempt to make assistant professors their minions: underlings who can write their grant proposals, papers, and other things the senior faculty might not care to do, or activities they are no longer competent to do since science has long passed them by. *At any level, good mentors always make their trainees increasingly more independent and not the other way around.* This, in fact, is the key outcome of good mentoring. I believe that it is only when scientists or engineers are truly independent that they can become valuable colleagues and collaborators. Again, assistant professors should choose their mentors wisely. Ask the department chair about which faculty members in your subfield are the very best mentors. Talk with faculty members who were mentored by the candidate offered by your chair. Better still, ask if you can be assigned a team of mentors. With a team, you can better manage the one you might discover who seems hungry for a minion.

Laws of Herman

Irving P. Herman is a physics professor at Columbia University, who offered advice to graduate students through an article that appeared in *Nature* (Herman 2007).

While he states that these "laws" are "slightly exaggerated" I think that they are the kind of advice a good mentor gives. In addition, he offers his advice in a humorous and disarming fashion.

1. Your vacation begins after you defend your thesis.
2. In research, what matters is what is right, not who is right.
3. In research and other matters, your adviser is always right, most of the time.
4. Act as if your adviser is always right, almost all the time.
5. If you think you are right and you are able to convince your adviser, your adviser will be very happy.
6. Your productivity varies as (effective productive time spent per day)^{1,000}.
7. Your productivity also varies as 1/(your delay in analysing acquired data)^{1,000}.
8. Take data today as if you know that your equipment will break tomorrow.
9. If you would be unhappy to lose your data, make a permanent back-up copy of them within five minutes of acquiring them.
10. Your adviser expects your productivity to be low initially and then to be above threshold after a year or so.
11. You must become a bigger expert in your thesis area than your adviser.
12. When you cooperate, your adviser's blood pressure will do down a bit.
13. When you don't cooperate, your adviser's blood pressure either goes up a bit or it goes down to zero.
14. Usually, only when you can publish your results are they good enough to be part of your thesis.
15. The higher the quality, first, and quantity, second, of your publishable work, the better your thesis.
16. Remember, it's your thesis. You (!) need to do it.
17. Your adviser wants you to become famous, so that he/she can finally become famous.
18. Your adviser wants to write the best letter of recommendation for you possible.
19. Whatever is best for you is best for your adviser.
20. Whatever is best for your adviser is best for you.

The Laws of Jason: mentorship from a postdoc's perspective (contributed by former graduate student and postdoc Jason Abercrombie)

Top 10 characteristics of a good mentor

Finding a good mentor in science can be difficult because the most important characteristics of a mentor have more to do with character and personal integrity than immediately apparent qualities such as verbally demonstrated knowledge and publication record. Graduate school can be an arduous path for not only your development as a scientist, but also for the refining of your interpersonal communication skills and character. If you pay attention, you will be constantly learning from people who have been around the block many more times than you and have learned and mastered, forgotten, or never learned some of the more important life lessons. When deciding on a lab to work in, it's a good idea to ask the current students and postdocs whether the professor demonstrates these qualities (good or bad).

1. **Understanding of individual strengths, weaknesses, and unique talents.** A good mentor knows that each graduate student has his or her own specialised tool belt of strengths that he or she can bring to the lab, while also recognising weaknesses. Good mentors will help to identify those strengths and unique talents while also helping the student improve in areas where skills are lacking.
2. **Clarity.** It is important as a graduate student that you know early on exactly what is expected of you in terms of work hours, specific duties, and number of publications required for the successful completion of your degree. A good mentor will clearly communicate these expectations at the start of the graduate program. The mentor will help you create a reasonable timeline in order to sufficiently complete those goals.
3. **Good sense of humour.** A good sense of humour is critical for a mentor in science. A mentor who has a good sense of humour makes the workplace a more pleasant place to be and has better relationships with his or her students. Learning to develop a sense of humour if you don't have one is equally important as you interact and develop collaborations with other scientists.
4. **Approachability and a good listener.** When you consider the seemingly infinite number of questions you'll be asking your mentor during your course of study, a good mentor will always be approachable to help answer those questions and provide you with guidance. The better listener your major professor turns out to be, the more fruitful your research will become. If he or she truly listens to the problems you are having in your research, then the most effective solutions can be quickly attained, and everyone wins!

5. **Capacity for correction.** As a student of science, you are going to be wrong a lot! But whether mistakes are the result of poor experimental technique or creating interpersonal drama within the lab, a good mentor has the ability to recognise these errors and correct both bad science and bad behaviour.
6. **Diplomacy.** A good mentor has to be diplomatic in order to effectively solve problems and deal with employees with conflicting personalities and behaviours. Diplomacy also is critical for fostering collaborations with other scientists. Diplomacy is an important life skill to develop over the course of your career as you encounter conflicts and disagreements.
7. **Organisation.** If, during your visit to a prospective professor's office, you feel like you've just entered "ground zero," you might want to reconsider. An organised person can foster organisation and productivity. The messy person is usually a mess.
8. **Leadership and vision.** A good mentor must provide leadership for the lab. Leadership without micromanagement allows graduate students and postdocs to make their own way without excessive floundering. A good mentor also realises that science is a process and that we "stand on the shoulders of giants" (Isaac Newton). Good mentors always see beyond the struggles of scientific discovery to the potential that science can benefit society. This vision enables the professor to see the value in their research.
9. **Forgiveness and patience.** The masters or PhD experience can be quite stressful and can result in harsh behaviour by the student. A good mentor recognises the potential in their students, and forgives them when stress brings out the worst. He or she understands that graduate training is not easy. The mentor can separate behaviour from a person's integral being. Few things in life are more rewarding than the results of patience. A mentor who has a lot of patience has wisdom. A good mentor recognises the inexperience of his or her students, and is patient during the course of their development as scientists. Patience and forgiveness go together.
10. **Communication.** A science professor is usually writing grants and manuscripts, and attending national and international scientific meetings. Therefore he or she must be a good communicator. A proven track record of publications in relatively high-impact journals and a high attendance record at scientific meetings is a good reflection of the professor's communication skills.

Top characteristics of a bad mentor

1. **Anger.** If a professor has a reputation of being easily angered, or is known as a "hothead," this can make your life difficult and create an anxiety-filled work environment. People burdened with unresolved anger issues generally don't make good mentors. Be careful not to become like this.

2. **Greed.** When research is driven by greed, and not science, bad things usually happen (e.g. falsification of data). By nature, science is an altruistic enterprise.
3. **Apathy.** When hard work and diligence goes unappreciated or unnoticed, it can be perceived as apathy or the expression of a professor's lack of interest in what their students are doing or struggling through during their research. A bad mentor is exclusively focused on his or her own pursuits, and is disinterested in the struggles and problems of his or her students.
4. **Condescension.** If someone is being condescending, it gives the impression that an individual has an intrinsic feeling of higher self-worth more than someone else. This is a bad trait for anyone to express, especially a mentor. Condescending individuals cannot mentor students because their students will always feel a lack of respect and dignity.
5. **Arrogance.** There is a fine line between confidence and arrogance, and often, people in science cross that line routinely; especially apparent in the arena of battling egos. However, someone with a reputation of being arrogant does not make a good mentor. Nobody likes associating with arrogant people.
6. **Complaining.** Constant complaining, grumbling, passive-aggressive behaviour, sour grapes, and the like are antithetical to being an effective mentor. If all you're doing is grumbling about how all the experiments are failing, why grants are being awarded to someone less worthy, blaming your circumstances on other people, and whining, then it is a reflection of your own poor performance.
7. **Use of fear or intimidation to produce results.** Although "slavedrivers" may sometimes do well in science and lead very "productive" labs, they are not going to express sympathy when you need to take some time off or have some life outside of the lab. During the process of completing a PhD, there will be many times when your research is consuming your life (and your thoughts) and having a PI breathing down your neck demanding more and more data is enough to make anyone miserable.
8. **Absenteeism.** The absentee mentor can't mentor. Some PIs take long and frequent vacations, sabbaticals, and any other excuse than to be at the university; they can become out of touch with students' programmes. A derivation from the absentee mentor is the mentor who comes to work every day but takes seemingly never-ending coffee and lunch breaks. While networking is needed in science, hours upon hours with colleagues each day socialising sets a bad example for students.
9. **Workaholism.** The opposite of the absentee mentor would seem to be good, but in fact, the workaholic mentor can be worse than the absentee mentor. This mentor lives only for work and science and believes that you should be the same way. Not only is it unhealthy, but fewer scientists these days find it sustainable.

How to train your mentor

Let's say that you want your mentor to be more helpful. Is it possible to improve your mentor's ability to mentor and therefore improve your own training? Let's say you want your mentor to be more responsive to your needs. It is hard enough to change our own behaviour and personality traits, so don't even try to change your mentor's. Instead, let's assume that all mentors' innate desire is to put their best efforts toward the directions of research so that their people can land the best results. So, one way to make a good mentor is to be a good student – see the Laws of Herman. I think that most faculty members instinctively want to "clone" themselves; i.e., produce a scholar who will follow in their own footsteps in the same field and be a faculty member at a research university. I know that's how I imagine students' destinies when they walk through my office door for the first time. I think, is this "the one"? The one who will do bigger and greater things in my field than me? Well, that approach is neither practical nor realistic, but one way to train your mentor is to allow the PI to think that the student wants to follow in the PI's footsteps. I'll admit this trick is somewhat disingenuous and I hesitate to mention it here because it could be construed as unethical, but the student really does want to learn everything the mentor knows and more, so this approach can be practical and effective. Students also have to be assertive to get good mentoring. Ask the mentor questions about the profession and other options after the PhD is earned. Maybe certain kinds of stress the mentor assumes is not for you, but there's no reason to lay all your cards on the table about career preferences and goals in every detail. The most important thing is to strive for excellence and that will seize the attention of your PI, giving you credibility and leverage to obtain what you need with regards to mentoring. I also appreciate students who keep me informed and want to dialogue about all their positive results in the lab. I appreciate the ones who ask me questions, come up with their own ideas and experiments and expect mentoring. Kearns and Gardiner (2011) conduct workshops to help students obtain better mentoring. They point out that the proactive students who make the point to regularly meet with their mentor with a specific agenda are happier with their mentoring process. Personally, all of these things motivate me to be a more responsive mentor. In science, the shrinking violets are never rewarded. Don't be shy with your mentor and expect to be mentored!

Case study: Choosing between two possible mentors

Joel has just finished his masters degree and is now looking at various universities and mentor candidates for his PhD. His masters degree was in the field of physiology and now he is looking towards adding mathematical and computational approaches to his repertoire. While Joel didn't publish any papers from his masters work from Comprehensive State University, he learned useful lab techniques, has high GRE standard test scores, especially in the quantitative portion of the exam, and shows good promise as a researcher. His masters lab was small and in a department where the masters degree was the highest one offered. Joel was a teaching assistant and is comfortable teaching and working in a small lab.

He visits two professors at Big Private University, a school where his masters professor recommended for further study in computational biology. The first (One) professor he visits does exactly what Joel wants, but has a small lab with little funding. He is friends with Joel's masters professor. Professor One says that Joel is a good candidate for a teaching assistantship, and already has a project exactly planned for a new student that would be perfect for him. He shows Joel an old grant proposal. Professor One says that while the proposal was not funded during the first two submissions, he still believes it is sound and a solid contribution. He plans on submitting it again: "Third time's a charm," he says. He also presents Joel with a schedule of month-to-month milestones and says that he will help Joel to stay on task by working side-by-side with him each day. He will also need monthly reports from Joel. Professor One's motto is, "Leave nothing to chance." Professor One has not published extensively, but his papers are sound – typically he is the first author, and proudly tells Joel that he is a perfectionist and will spend several weeks poring over each section of Joel's writing. One expects each sentence to be perfect. His graduate students are typically co-authors. Joel likes the thought of the One-on-one attention. In addition, Professor One reminds Joel a lot of his masters professor, but with a better publication record. Joel's CSU professor didn't publish much and Joel was looking forward to really seeing how publishing is done. While Professor One does not show Joel his laboratory during his visit, he says that the computational facilities across campus are exceptional and that the campus supercomputer can be used for Joel's project. In addition, Professor One guarantees Joel will be able to finish his PhD dissertation in just three years and be directly competitive for a faculty position in the field. At the end of their meeting, Professor One asks Joel not to talk with anyone about the project he has described that could be lead to Joel's dissertation. One says that it is important to keep the project secret because someone else might steal the idea and get funding with it instead. Joel agrees to keep the project secret.

Joel then visits with a second (Two) professor during his day-long tour of BPU, but she does not have as much time to spend with Joel as One did. She introduces him to several graduate students and postdocs in his lab instead of talking with Joel for his entire visit. Professor Two has several NIH grants and is working in both physiology and mathematical modelling. The project that she has for Joel is part of one grant, in which the lead work will be done by a postdoc, Dr. Quattro, who has extensive experience in molecular modelling. Dr. Quattro seems to be a nice person who is very focused on science. Both Quattro and Two say that Joel should get between three and four first author papers if he works hard and shows initiative. Neither Professor Two or Dr. Quattro is quite as warm as Professor One, but they both are well-respected and liked by the other people in the lab that Joel meets. Dr. Quattro also shows Joel the computer cluster where most of the modelling work is done and, in disagreement to what Professor One said, Quattro claims that the supercomputer across campus is not readily accessible and would also not be needed for any project that Joel would likely do. This apparent disagreement about equipment accessibility and need is somewhat confusing to Joel and he thinks that not learning to use the supercomputer would be a disappointment. Professor Two's lab is also quite a bit larger and more productive than the small lab Joel did his masters work in, but it is also more intimidating. Professor Two says that Joel would be on a research assistantship and would work closely with Dr. Quattro. While Joel wouldn't be working directly with Professor Two in any aspects of the research, she has an open door policy where people can walk in with any questions. She also says that it might take Joel longer than three years to do his PhD in that the time for data collection and writing the papers for publication is sometimes lengthy and unpredictable. She did say, however, that her students have finished their PhDs between three and five years, but that she cannot guarantee that Joel will ever complete the PhD, because much of the success or failure is up to each individual student.

1. Joel is comfortable with Professor One's hands-on approach, which is more like his previous mentor. How important is this feeling of comfort and familiarity in making his decision?
2. How important is it to work on a funded grant or at least, a line of research that seems fundable?
3. What should Joel think about Professor One's desire to keep the project secret? Do ideas often get stolen in science?
4. Should Joel worry much about the prospect of working more with Dr. Quattro instead of Professor Two on his PhD? In contrast, he'd be working very closely with Professor One if he chooses his lab.
5. How should Joel feel about the different information he receives about the supercomputer? How should that affect his decision?
6. How should Joel feel about One's guarantee that Joel would complete his PhD in three years versus the absence of any guarantees with Two?
7. Which lab should Joel join and why?

Choosing the right research project: the new graduate student's dilemma

This fictional case study is provided courtesy of graduate student Jonathan Willis.

Karen Sparks joined Dr. Amie Leavens' lab as a new Master's student in a marine science department. Karen aspires to eventually take a PhD in marine biology and wants to choose a research topic to eventually position her for a tenure track position at a research university. Karen has just started learning molecular biology.

Dr. Leavens has recently received tenure and won two grants. The first is an applied project on environmental toxicology of copepods and the second is more of a basic study of the molecular interactions within the copepod and how gene expression changes because of environmental changes. At the end of Karen's first semester, Dr. Leavens approaches her wanting to know which of the grants she prefers to work on and asks her to set up her graduate committee.

Dr. Leavens believes that a masters study should be very focused and straightforward. She suggests that Karen could simply execute a portion of the more applied toxicology grant as written. The molecular study is far more in depth, more uncertain as to the results, and would involve more time and some additional experiments that would require some methodology to be worked out. Dr. Leavens also mentions that she thinks that perhaps one large or two smaller papers will come from the toxicology work, but at least two- and possibly three bigger and more important papers could result from the molecular work.

Dr. Leavens has funding for two years of toxicology studies and three years for the molecular research. She feels strongly that more funding could be attained for future molecular work based on very early preliminary data. There is a post-doc in Dr. Leavens' lab whom will perform molecular biology on a different goal of the grant. If Karen chooses the molecular route, she will work closely with the postdoc. If she chooses the toxicology project, it would be solely up to Karen to perform the research. Dr. Leavens tells Karen that based on her background she feels confident that she can achieve goals pursuing either project, but the molecular project will require more creative experimental design and more work than the other. She tells Karen that there is more certainty of completing the toxicology research in two years but cannot predict what will be the actual timeline for the molecular study given the circumstances.

1. What factors should Karen weigh up when making her final decision for her graduate program?
2. For the toxicology project she will be a more independent worker, but not play a creative role in experimental methods, since those have been finalised in the grant proposal. How much should that matter to Karen?
3. A tradition doctorate program in biology ranges from three to five years, depending whether a masters degree is earned, how fast the research becomes productive, and other factors. If Karen pursues the doctorate with the molecular study she is not guaranteed funding after the third year. How important is not having guaranteed funding after three years to the student?

Judge yourself *redux*

✓ How good a judge of character and environments are you? Can you "size-up" people? How rapidly?
✓ Are you an "easy sell" or able to resist promises that are too good to be true?
✓ What are your most important features in a mentor and laboratory? How does your personality and style play into your choices?
✓ How do you feel about mentor accessibility? Do you prefer to walk in and chat when someone's door is open or would you rather make an appointment? Are you a texter or an emailer? What is the mentor's communication mode and style?
✓ Are you a trusting person? Do people tend to find you trustworthy in return?

Like most of these segments there are no right and wrong answers. I think it is really important to "know thyself" so that you can match your personality and style with that of a mentor. It is also important to gather information that is trustworthy. A mentor-mentee relationship is like a temporary marriage.

Judge yourself *redux*

✓ How aggressive or assertive are you? Do you feel like you take control of situations?
✓ Are you a good judge of character?
✓ Do you have trusted friends and colleagues whose opinions you trust?
✓ Do you listen to advice?

This series is much like the last one. It is very important to feel some sort of control of the mentor selection opportunity (while it is mutually selective). It is also important to listen to advice. I got better at this as I progressed through my career, but I wish that I'd sought even more advice earlier.

Summary

Due diligence in picking a lab and a faculty mentor is crucial as a springboard to a successful career in science. It is important to like and respect your mentor, therefore it is important to search for likeable and productive mentor candidates who possess traits that command respect. A friend of mine has four criteria, to paraphrase, for selecting a mentor: 1) lab with money, 2) lab with necessary equipment; 3) nice mentor; and 4) productive mentor. Don't be fooled by words – check out potential mentors' claims by actions and their research record. Graduate student and postdoc trainees must be assertive to receive the mentoring they deserve. Matching good student and good mentor who work well together is a huge benefit to science.

Chapter 5

Becoming the Perfect Mentor

ABOUT THIS CHAPTER

- Good mentoring of junior scientists helps them to become successful senior scientists.
- Imparting skills to foster independence is the goal of mentoring the next generation of scientists.
- Teaching trainees how to publish results and acquire funding are two vital skills needed for success.

After all the schooling and postdoctoral work is completed, when you have your own lab, what kind of mentor will you be? The best mentors allow their students and employees to fulfil their own personal and professional goals while advancing science. One of the first questions I ask the visiting student or postdoc candidate is, "what do you want to be when you grow up?" The answer reveals the person's career goals. In addition, how they approach this question can be telling. It is rare that anyone has long range goals clearly delineated, and therefore, mentors have a huge opportunity to help clarify fuzzy goals and help develop milestones to reach them. As mentees mature and grow as scientists, the mentor has the privilege of helping them find and follow their dreams. Mentoring comes in stages, and the kind of mentor you are early in your career might change as your career advances, and that's normal. It seems to me that as I get older, enabling others' success becomes the primary mode through which I can secure my own success in science. Just as mentoring styles and goals change during career stages, each trainee has individual needs. Some scientists-in-training need significant guidance, especially in their early training. Others simply need to be pointed in the right direction and have their path cleared. The custom mentoring of diverse trainees is often tricky but is rewarding.

Grants and contracts are a prerequisite to productive science

The professional aspects of becoming a good scientist always return back to productivity – publications and grants to fund the research that leads to discoveries

Research Ethics for Scientists: A Companion for Students, First Edition. C. Neal Stewart Jr.

and publications. If a graduate student or postdoc wants only to teach, then grantsmanship would appear to be less important, but what if there is a change of heart and they also want to pursue research? Unless they go into industry, then grantsmanship is important. I guarantee that a teaching college will find evidence of grant-getting attractive, because more and more administrators in teaching colleges view research and teaching grants as vital to expanding and improving their programs. Therefore, mentoring is multidimensional, even in the simplest of desired outputs.

I argue that becoming grant-savvy and a grant-getter has value no matter what the career goal. I frequently ask graduate students and postdocs to write portions of grant proposals where I am the PI. After that, I'll request that they craft their own proposals for either me to submit or, if they wish, they can serve as PIs and submit them on their own. It is often best if I remain as the PI on the grant since many universities do not allow graduate or postdocs to be the PI on a regular full-sized grant. If I am the PI, ethics prescribe that I confer control of the science to the postdocs who wrote the proposal while monitoring progress. There can be the opportunity to transfer the grant to them outright should they transition to positions appropriate for them to be the PI for the grant. I can also help them become familiar with the budgetary and paperwork aspects of proposals and grants after they are funded. As a mentor, I can also help them manage the people aspects of research.

Grant proposals are not easily funded, in part because there is a learning curve in writing a fundable plan. The research also must be modern and thorough. For most scientists, it usually takes trial and error before winning. Even for experienced PIs, there is difficulty in acquiring funding, which is always statistically, improbable. But the probabilities fall to zero if a proposal goes unwritten or unsubmitted. I like to think of practicing writing grant proposals to be a relatively low risk activity for graduate students and postdocs since they typically don't absolutely have to win their own funding for my success as scientists at very early career stages. In most cases, they are funded for their existing positions in another grant. Therefore, it is a win-win situation if their proposals are funded but not absolutely necessary to be deemed successful postdocs, for example. This situation changes drastically when the tenure clock begins to tick. An assistant professor needs to be successful in the grants-world to earn tenure at many institutions. I prefer for my postdoc trainees to already have "paid their dues" in grant-rejection-land prior to when it actually counts. Naturally, they will still experience grant proposal rejection as assistant professors, but chances are they'll also experience a higher frequency of success.

Judge yourself

- ✓ Do you enjoy planning experiments and asking for resources to do research? Do you enjoy teaching these skills to trainees?
- ✓ How much do you believe in your abilities to find funding for your research and teach others how to do the same? Can you imbue confidence in others?

✓ Are you a trusting person? Can you rely on others to succeed? Are you devastated if your trainees fail? How do you deal with success? And cope with failure?

Publications are the fruit of research

One huge role of mentors is to make publications happen. The primary route is through the publication of theses, dissertations and postdoc research. This means allowing and expecting trainees to take control of a study and the resulting manuscripts; own it, write it, be the first author. It also means encouraging teamwork and coauthorship. It is bad mentoring to allow part of a thesis to languish unpublished. But sometimes this still happens in my group and I get mad at myself whenever it does. The best scenario is for students to submit the papers before they write their dissertations. After a student leaves the lab, a professor's leverage is lower and passive trainees might never write-up their results and publish. On rare occasions papers go unwritten, unless the professor does it; students sometimes leave, moving on to other pursuits and forget about their unwritten manuscripts. While professor-written erstwhile student papers might help science (better than the data going unpublished), this practice defeats a major purpose of mentoring: trainees mastering all aspects of manuscript preparation, peer-review, and revision. Under the umbrella of the mentor, this mastery optimally occurs when there is the option to fail with little risk or penalty; this situation changes when the tenure clock is ticking. Trainees must gain skills of scientific independence if they are to become fruitful contributing senior scientists.

My first priority as a professor is to encourage manuscripts to be drafted, vetted by all the co-authors, finalised, and then submitted as quickly as possible. I don't understand why a professor will sit on a manuscript for weeks and months, but I see this happening around me all too often. I hear their students and postdocs complain bitterly of this situation. I don't think their mentors appreciate the depth of their trainees loathing when their PIs procrastinate and badly manage research. I can understand some reasonable delay in submitting a paper if a grant proposal is taking precedence, but students and postdocs need timely publications for them to realise their dreams, which typically includes graduation and career advancement.

On a personal level

Good mentors also realise that their mentees are more than science robots. They are people with feelings and personal goals and dreams. I've listened to students talk about their friendships, love lives, parental aspirations, health problems, religious beliefs, politics, philosophies and exercise regimes. When asked, I've given advice on what classes to take, what cars to buy (or avoid), who is a trustworthy collaborator, and the difference between ethics and ethnics. Unprovoked I've traded stories and had arguments about which barbeque, beer, and football teams are the best, renting vs buying, boats, musical instruments, table tennis and why christopherwalkeniswatchingyoupee.com is kind of creepy and amazing at the same time. The best mentors continually listen to their trainees and give advice when needed, and not the other way around. Mentors can sense when their trainees' aspirations change. Personal situations certainly affect professional performance, therefore mentoring must be holistic in nature. That said trainees' personal lives and boundaries must be respected. No one likes a nosy boss, but everyone wants someone to care about them on a personal level. There is a fine line to be observed here. And in no circumstances do trainees want to hear about the boss's personal troubles. It is ok for mentors to lament about their recent manuscript being rejected but not ok to talk about their domestic arguments.

Graduate students and postdocs need freedom to do their desired experiments and follow their noses that will lead to good publications. I stopped doing lab work more than a decade ago. Therefore, since my two hands are no longer performing experiments, the science that comes out of my lab is the only science I can claim, albeit vicariously. I have no backup plan for collecting data and generating papers – I'm completely dependent on my trainees for my success. The mentor is more like a coach than a player, so let's end this section with quotes from the late basketball coach John Wooden. "A coach is someone who can give correction without causing resentment." And, "You can't live a perfect day without doing something for someone who will never be able to repay you." These words paint the picture of a perfect mentor.

Judge yourself

✓ How much do you care about people? Specifically, how much do you care about people on your team?

✓ Are you empathetic? Are you aware of lines that you should not cross that would otherwise make people intimidated or uncomfortable?

Common and predictable mistakes scientist make at key stages in their training and careers and how being a good mentor can make improvements

Graduate students

Many graduate students don't immediately make the transition from being an undergraduate and having knowledge spoon-fed to them to being aggressive when obtaining knowledge for themselves. The U.S. system, which requires certain additional coursework during the masters and PhD degrees, doesn't help this situation much, in my opinion. Requiring formal courses is not all bad, but if graduate students are taught in the style as when they were undergraduates, this passive learning style is reinforced. If students are at their desks "studying" more than they are in the research lab doing experiments, they are not becoming independent scientists. Also, nearly all students make the mistake of not writing enough, and when they do write, it is often not in a style conducive to science publications. Many science students are not good writers, but if they don't become comfortable writing and speaking in English, their careers will never be fully launched or their scientific potential realised.

Graduate students can become mentors to undergraduate students performing research in the lab and also to entering graduate students. This mentoring not only helps the trainees, but also aids in graduate students making the transition to becoming a professional scientist. It is well-known that teaching helps the teacher gain a deeper understanding of the topic being taught. Indeed, students respond well to mentors who are often not far beyond their years and experiences.

Postdoctoral fellows

The common early mistake of a postdoc is to dwell a bit too long and too seriously on the glories of finally obtaining the PhD and having the title of doctor. Postdocs don't realise soon enough that they are really just getting started in research and have a lot to learn. Suddenly growing a big ego, strutting around the lab and barking out orders is not the way to engender learning or cooperation from those around you. Another common problem arises when postdocs do not grow sufficiently independent as researchers in their studentships, and are not quite ready to be independent postdocs. I've heard my lab manager utter the phrase, "you're a postdoc now" in an attempt to inspire independence. There are many times when this phrase (which can be sung to the tune "You're in the Army Now") doesn't translate to instant independence, and maybe for good reason. When postdocs change fields and learn several new topics while simultaneously getting used to new surroundings, it might take some time for them to feel independent and productive. Nonetheless, postdocs must challenge themselves to gain a sense of independence and competence, because that is what sorts out effective scientists from the crowd. At the end of the postdoctoral period, scientists should have had the opportunity to learn from their own mistakes in mentoring, grant proposal writing and publication so they don't have to make the same mistakes during their assistant professor years while their tenure clock is ticking.

They can help themselves by trying new tasks, including the sub-mentoring of graduate students and undergrads – really taking charge of a project. Having recently completed dissertations, preliminary exams, and the angst of getting their degrees, postdocs can be of great comfort to the students they are mentoring. Mentoring also helps postdocs see beyond themselves and towards service to their labmates and greater science.

Assistant professors

Some assistant professors who have not had sufficient and effective postdoctoral training make predictable mistakes in writing grant proposals and mentoring. Being able to write effective grant proposals is crucial to getting tenure in most universities. The ability of assistant professors to win grants is actually more important than their doing experiments themselves in the lab. And since assistant professors might be decreasingly present in the lab, they must learn to recruit and retain graduate students and other trainees.

In fact, mentoring mistakes (beware of being a new hire's first grad student!) are bound to occur as the assistant professor is learning the ropes of mentoring. A big mistake made just after recruiting that first graduate student is not letting go of the research reins. As we know, graduate students must own their own projects. Don't micromanage. Also, don't be too buddy-buddy – you're the professor now! Collegiality is good. Sleeping with students is bad. Another mistake assistant professors often make is not having a balanced life. By worrying so much about making tenure, they don't have a life and actually become less effective and tenurable than if they pursued some interests outside the lab. Other mistakes include teaching too many classes, taking too much time to prepare for teaching, and engaging in excessive university and professional service. In almost all universities, tenure is awarded primarily on research productivity (i.e., peer-reviewed papers and how much they are cited) and grant awards.

Associate professors

Associate professors shouldn't take long naps and vacations after they are awarded tenure. Quitting research for even a short period of time results in a higher probability the research program will cease to be competent and competitive in modern research. When it ceases to be a fun challenge, it is best to leave research. One mistake that associate professors make, especially if they decrease their research productivity, is to start down the vortex of too much university and professional "service" and teaching and not enough recruiting and mentoring of trainees. Feeling guilty about the lack of research productivity and also, experiencing some immediate rewards from running big professional society-, department-, and university committees, serving on the faculty senate, and doing more teaching, the associate professor is now poised to be even less competitive in research and less likely to be an attractive mentor for graduate students and postdocs. Like tenure, being promoted to full professor is usually based upon research production and whether the candidate has an international reputation in his or her field. Therefore, serving in more editorial roles, organising conferences, serving on grant panels, and research-oriented service is a more effective path to becoming a full professor than chairing a big faculty senate committee (leave that to the full professors who are close to the end of their careers).

Full professors

The proverbial "deadwood" at universities is almost always full professors waiting to retire (albeit sometimes waiting decades!). With post-tenure review becoming more common, tenure-for-life is no longer a death-and-taxes certainty in academia. It seems to me that if someone chose to pursue science because of the intense interest in research, then why should that necessarily change with age? Some full professors lose the competitive edge and do not stay current in the field. It does not take long for science to accelerate past static competence and comfort into a realm of unfamiliarity. In finding that they are no longer competitive for grants, full professors sometimes quit doing research altogether. Sure, increased university service is rewarding and someone has to do it, but keeping current in science is more fun. And science is more fun than coming in late, going home early, and drinking coffee and shooting the breeze with the like-minded in between. In addition, the best full professors are efficient enough to have fun at more than one thing at a time.

It is good for the full professor with tenure to remember the fun of doing science. Sabbatical programmes, outside training, and rekindling research in hands-on laboratory experiments can be helpful to jump-start research. Doing research in a very different field than the one in which a faculty member gained a reputation can also be scintillating. It is a big mistake to become a member of the deadwood coffee club. As the reticently retired American football coach Bobby Bowden (he retired at age 80) said, "After you retire, there is only one big event left." In this stage of their careers, full professors should now know enough to be expert mentors. Experience is a terrible thing to waste. Science should benefit the most from full professors sharing knowledge with their younger trainees. Full professors should be inspirational. See the box on Bob Langer for an excellent example.

Mentorship of many trainees: the story of Robert Langer

Source: Bachrach Photography.

Robert (Bob) Langer has a gigantic lab in Chemical Engineering at MIT. Helen Pearson, a writer for *Nature* followed him around one day to learn why and how he was so productive (Pearson 2009). Dr. Langer is one of 14 Institute Professors at MIT, the highest honour that MIT bestows, has published over 1100 papers, and is the inventor of over 760 issued or patents pending. He is the most cited engineer in history and has won over 180 major awards. He is one of few people ever to be elected as a member to all three National Academies (Engineering, Medicine and Sciences). Given his extraordinary productivity and genius, I was even more impressed by why he does what he does: he says wants to help people and make them happy – so he said in the *Nature* article. Therefore, it seemed to me that Bob Langer might be the perfect person for me to ask a few questions about mentoring. Sure enough, he answered my request email after just a few minutes and agreed to an interview for this book.

Neal: I read in *Nature* that you have 80 or so people in your lab at MIT, which would make it the largest research lab at, arguably, the best scientific research institute in the world – maybe the largest research lab in the world. The size seems overwhelming to me. Could you tell us something about the breakdown – the number of postdocs, graduate students, etc., and what is the most overwhelming part of managing such a big group?

Bob: Currently, I have probably around 25 graduate students, maybe 45 visiting scientists and postdoctoral fellows, 8 technicians, and 5 office staff. At any time there are 30–50 undergraduates in the lab doing research (called UROPS at MIT). It's impossible to give you an exact breakdown. The most overwhelming part? You know, it's funny – I don't really find it that hard. Growing over time might be hard, I suppose, but maintaining a steady state is not that hard. You just develop a style that works. I have a really good office staff and there are senior postdocs – it just works itself.

Neal: You must be a good manager of your time and resources. What is your "secret"? What advice could you give assistant professors embarking on a career in academic research science?

Bob: I think I am good at delegating. I don't know that I have any secret. Obviously, I work hard... lots of people work hard. I've been a pretty good delegator. I don't think it's important that you have a big or small group. The advice I always give people is that it is important to work on projects that you really believe in; projects that you feel will make a difference. It's important to take risks and work on really important high impact projects. Be good to your students and postdocs – help them in any way that you can. It's really important to obtain solid funding. These are some of the priorities that I tell people to do.

Neal: What is your mentoring style? How does mentoring differ among people in your lab? That is, do you mentor graduate students differently from undergrads? Postdocs differently from grad students?

Bob: To me, I consider that people are at different stages of life. Until the time you do research, you spend your life answering questions people ask of you. If you're in high school or in college courses you're judged by how well you do on tests. Even as an undergraduate and sometimes graduate student working on research projects that other people come up with – you are judged on how well you do on them. My goals for undergraduates are to learn good lab skills and get excited about research. For grad students – I want them to begin to make the transition from giving answers to asking questions. And for postdocs, even more of that.

Neal: So it seems that you encourage your students and postdocs to "own" their projects. Do they take the projects with them when they leave your lab, or do you keep part of the projects? What is your policy?

Bob: I do tell people it is fine to take their projects. However, it is important for their sakes that if they take aspects of a project that they do something different with it. If the NIH or their new faculty think they're doing something that just duplicates their postdoc project, they won't look favourably on them. I think it's best for people to do things that build on what they've done. But if someone wants to take their project, we have so many things going on that it's ok. But I want to encourage people to think big and build on what they've done.

Neal: Do you have an open door policy? If not, how does your staff access you?

Bob: I used to have an open door policy, but now I have an open book policy. People just put their names in my book or tell my secretary, or they also email me and I try to get back to them quickly.

Neal: I did notice that you are a rapid email responder. How do you manage all of the information you are barraged by? From emails to the literature and things in between?

Bob: I don't like things to build up and I like really try to finish everything everyday so they don't build up. Otherwise I get stressed and if I get things done I don't feel stressed. I feel like people appreciate it when I get back to them on email and that means a lot to me. Like the article in *Nature* said, I like to make people happy. I'm the kind of worker that'll keep going at it 'til I've finished the things I need to do. For example I'll do email while I'm walking. I do answer a lot of email.

Neal: What do you think is the most common mentoring mistake that faculty members make? The most damaging?

Bob: I think that faculty members have to walk the line. Some faculty members are too controlling from my standpoint, and maybe others are not [controlling] enough. Sometimes they tell people too much of what to do and then students don't get excited or don't learn the keys of what research really is or how exciting research can be. I think it's important to give students a certain amount of rope and to show them you're always there for them. I think the biggest mistake is oversupervising.

Neal: On the other hand, what is the best thing a mentor can do for his postdocs? His graduate students?

Bob: The best thing that you can do is to help them get really great ideas, and how to do research that has a huge impact – to show them the excitement of research.

Research can really be a wonderful thing. You want them to have good credentials, good papers, and good grants.

Neal: In an earlier answer, you used the phrase "walk the line." I think true artists and successful scientists are highly adaptable – that if Johnny Cash were starting out in music today he'd be poplar in some genre of music. And if past Nobel prize winners were starting out in science today that they'd contribute in modern science too. What is the mentor's role in nurturing adaptability and "genius," or do you think these are characteristics people are born with?

Bob: This is a really interesting question. I think it is some combination of people being born with talents and other things. But I think that mentoring plays a big role. I also think that circumstances can play a big role. Maybe you've read Malcolm Gladwell's book "*Outliers*." Circumstances about where you are in history can play a big role. Mentors can play a big role. For me, I had Judah Folkman as my mentor, and that played a big role. Mentoring certainly helped me. I don't think there's any accident that there's been a couple hundred people from my group who have gone on to be professors and another couple hundred people from my lab who are now entrepreneurs. I saw success from him [Folkman] and he was a good mentor for me. And I've liked to see people from my lab do well. They're like family to me.

Neal: I'd appreciate any other words of wisdom you might have for young scientists starting out in research.

Bob: Gee, I'm not sure. I think that goes back to what you asked before. I think that it's really important to pick challenging problems that have high impact and then be sure to raise enough money.

Mentoring advice from a Bob Langer trainee: Daniel Anderson

Dr. Daniel G. Anderson, Associate Professor, Massachusetts Institute of Technology.

Hearing Bob Langer discuss mentoring had a profound impact on me. Bob's mentoring philosophy provided confirmation that I was doing certain things right and challenged me to make improvements. Playing the skeptic led me to check out Bob Langer's story, so I also interviewed one of his long-term trainees, Daniel Anderson, who has been a research associate in the Langer group for about ten years and who is now an

Associate Professor at MIT. During this time, Daniel has sort of built a lab within a lab and tells me about his own mentoring style and what life is like in the Langer Lab at MIT.

Neal: Tell me about yourself. What is your background? How did you find yourself in Bob Langer's lab?

Daniel: My PhD research was in the field of molecular genetics studying enyzmes involved in DNA repair. My degree came from UC-Davis. I came to MIT as a postdoc and had the chance to stay and build a subgroup. I wanted to move toward more applied research and gene therapy. My subgroup consists of about 40 people: mainly postdocs doing research in the area of drug delivery.

Neal: When you interviewed with Bob prior to joining his lab, do you remember any outstanding moment or event that shaped your decision to join his group?

Daniel: When you visit a lab, you want to get a sense of the impact of the place and people. You ask yourself, "Is this where you want to be to get stuff done?" You want to get a sense from other people in the lab as to whether they are happy and productive. On the other hand, you also want to sense if people are moody and if there are challenges to being productive. I think that if people are happy, that more work is accomplished, and so you screen for hard-working people who can work together as a team. So, when I visited at MIT and had a conversation with Bob, he offered me a postdoc position. I was struck with the impact that his lab had on science. It was clear to me that the work that was getting done had a high impact. Although I knew about the impact before my visit, during my visit I got a more tangible connection.

Neal: What do you think is the most common mentoring mistake that faculty members make? The most damaging?

Daniel: The one that sticks out to me is when people are jerks; when they treat their lab people like dirt. That one might not be the most common, but it hurts people and productivity the most. You need to be sensitive to people. Some mentors want to motivate people to getting every drop of productivity out they can, but it seems to me that by spending more time calming people down, they can get motivated. Mentors should motivate by the impact the research can have. But you have to be sensitive to people and their needs. I'm more interested in people being happy and productive. Typically, if they are happy, they'll be productive. We want people who are self-motivated.

Neal: What is the best thing a mentor can do for his postdocs? His graduate students?

Daniel: Spend time with them and think about what their projects are and help them. Mentors should read what their trainees write and help them.

Neal: What have you learned from Bob about mentoring, science, or life in general that you would like to share?

Daniel: In running a big lab, you have to have enough money to do it. Resources go beyond salary that is needed to do experiments. I've learned that the social challenges of dealing with a large group of people is sometimes more difficult than the science; good interactions among people keep it healthy. Work hard and play hard. Good people come to us . . . finding money is a lot of work.

Case study: The case of the missing mentor

PhD student Mitch Mitchell is a first year student in the lab of Dr. Thomas DeBague, which focuses on applied microbial biochemistry. The particular subject area of the lab has several practical applications and plenty of career opportunities for PhD-level scientists. Mitch enjoys this aspect of DeBague's expertise and initially viewed working in his lab as a great opportunity, since it is one of the few labs in this particular area of research. Mitch's main problem is difficulty in scheduling meaningful mentoring time with his major professor. Meetings with Dr. DeBague don't occur for several reasons. The first is Dr. DeBague's heavy travel schedule; second, poor health; and third, when he is in the building, it seems that he is either drinking coffee and eating donuts in the breakroom or sequestered away behind his closed office door. A secondary, but related problem is that Dr. DeBague is not quick to respond to Mitch's emails or read and comment on Mitch's writing. Often, Mitch doesn't know if DeBague is in the office or even alive.

Another problem is that Mitch is the only graduate student in the lab and feels lonely. In fact, Mitch has few friends or social life and wants only to do science. He feels like he gets inadequate basic technical training in necessary biochemistry techniques, and virtually no mentoring. He wants a lot more direction for his PhD program. He did somewhat similar though simpler techniques for his masters degree, in which he was mentored by a young assistant professor, Dr. Fabé Energé, one of Dr. DeBague's first graduate students when Dr. DeBague was an assistant professor himself. In Mitch's masters lab, he got accustomed to an atmosphere of high levels of attention, and lots of day-to-day instructions and directions. Dr. Energé said that Dr. DeBague would be the perfect mentor for his PhD.

Mitch came to the conclusion that there must have been many changes in the intervening 16 years between Dr. Energé's and Mitch's tenure in Dr. DeBague's lab. When Mitch does manage to get Dr. DeBague's attention, DeBague compels Mitch to enter his office; DeBague then closes the door and subjects Mitch to seemingly endless stories about when Dr. DeBague was a graduate student and an assistant professor, how difficult his own graduate experience was, and how he has finally "made it" in science through his own ingenuity, energy, and hard work. Dr. DeBague is fond of saying, "I have tenure now so I can do whatever I want whenever I want to do it." After that cue, Mitch is typically then treated to travelogues about Dr. DeBague's latest vacation or conference. Mitch is very patient during this routine, perhaps too patient. Finally, whenever Mitch gets around to requesting professorial guidance, or asks about the proposal he gave him to read, Dr. DeBague finds something else that is more urgent. "I must now confer with my colleagues about an important project," DeBague will typically say. "Keep working hard and you'll be as successful as me."

Mitch does make modest progress in spite of being alone in the laboratory. He drafts his dissertation proposal early in his second semester as required. After two months, Dr. DeBague finally gives the proposal back to Mitch with few comments. "Not a half bad proposal, if I say so myself," DeBague proclaims. Mitch's graduate committee generally lauds his efforts, but points out key technical flaws and scientific gaps in his proposal. They also point out that Mitch seems tentative during much of his proposal presentation. Mitch thanks his committee and vows to do a better job, but leaves the meeting feeling even more dejected and desperate.

Mitch feels that he is at a crossroads as a graduate student. He has tried repeatedly to engage his major professor, but seems to be stuck in an endless loop that is never resolved or productive. What should he do?

Questions:

1. Should Mitch try a different strategy to get good mentoring from his major professor? If you think this is a good solution, how might he approach it? Is Mitch blameless in his situation?

2. What is the chance of success should Mitch decide to continue to go it alone as he has in hopes that Dr. DeBague is right? Do you believe the assertion that Mitch will be successful if he works hard? After all, isn't developing scientific independence a key outcome during the PhD process? What are the advantages and disadvantages of taking this approach? Is Mitch working towards independence?

3. Should Mitch leave the university or simply leave his department? Or should he stay in his department but switch labs? Or something else? He's taken a class with another professor who is engaged in research that is different than that what Mitch originally wanted to do, but Mitch has talked extensively about his problems with the other professor's graduate students and has seen merits in the new area; he thinks he might be able to cultivate interests in this area. The other students confide that since it is a larger lab that they aren't alone in science, but also complain that their professor is very busy. However, they say they do see him and get directions and advice, but without the travelogue. Their professor also is responsive to email and reads and comments on their written projects – it only takes a few days to get back constructive comments. They encourage him to speak to their professor about switching labs. What are the advantages and disadvantages to switching labs and major professors? Would it place the alternative professor in a collegial bind to accept Mitch? Should Mitch change departments or universities to salvage his science career without Dr. DeBague?

4. Are there any other potential solutions?

Judge yourself *redux*

✓ Do you enjoy planning experiments and asking for resources to do research? Do you enjoy teaching these skills to trainees?
✓ How much do you believe in your abilities to find funding for your research and teach others how to do the same? Can you imbue confidence in others?
✓ Are you a trusting person? Can you rely on others to succeed? Are you devastated if your trainees fail? How do you deal with success? And cope with failure?

The first question really asks about the enjoyment of the role of the modern research scientist. This is what we do and I enjoy it very much. The second question deals with confidence in fulfilling the role of überscientist in your lab. The boss has to have a critical amount of mojo to spill out of his office and onto the crew. Winners know that they should win. Basketball scorers – even after they miss a shot – will shoot again and again because they believe they will score. The only systematic way to succeed is to fail. In learning science for yourself, there is a good amount of floundering around that must take place. I don't like to see people fail, but I do enjoy watching their growth.

Judge yourself *redux*

✓ How much do you care about people? Specifically, how much do you care about people on your team?
✓ Are you empathetic? Are you aware of lines that you should not cross that would otherwise make people intimidated or uncomfortable?

By nature I am probably a loner. Still, I want to be with my team and be there for my team so their dreams can come true. My own PhD-getting period was not easy for me, so I do have empathy. I've also witnessed tempests in a teapot when lines are crossed so I don't wish to replicate bad experiences I've observed and read about others.

Summary

Learning to mentor is a process that takes effort and courage to do well. Mentoring is dynamic and the roles and duties of mentors change throughout a person's career. Given the choice between being too controlling and less controlling, a mentor should choose the latter. A major goal in mentoring is to increase the independence of trainees. It is vital that mentors keep their excitement for science kindled so they can spread it on to their trainees.

Chapter 6

Research Misconduct: Fabricating Data

ABOUT THIS CHAPTER

- The most blatant of research misconduct is arguably the first F in FFP: fabrication of data.
- The basis of sound science is sound and honest data.
- Scientists justify data fabrication using a myriad of reasons, none of which are valid.
- Image fraud – using figures inappropriately to deceive – is a growing problem in science.

Science rises and falls with data and their analysis. I believe that most scientists play it straight-up: they want to know and report the truth. They collect data from well-designed experiments and report the data that tell the whole truth. Not all experiments work; neither are all data created equally. Scientists are mainly an objective and honest group of people: highly educated and with a sense of research ethics. But, research misconduct is disturbingly prevalent. It is visible in high profile cases in high profile journals. Therefore, the honest researcher often becomes invisible to the public while the infamous cheater becomes notorious thereby tainting the reputation of science. Invariably, when such cases are made public, it is the result of discovery of misdeed when a reviewer, editor, or another scientist in the field reports that data are missing, wrong, or simply fabricated. Of course, honest mistakes happen in every field of endeavour, and science is no exception. Honest, even substantial, mistakes in a paper can result in the paper being retracted by the authors. Papers that are found with falsified or fabricated data are also typically retracted, but with more dire consequences that accompany findings of research misconduct. I know colleagues who believe that I and some other scientists over-emphasise the problem of research misconduct, but I believe it is actually underestimated on the whole. As a letter-writer to *The Scientist* pointed out: "If such major journals as *Nature* and *Science* could be fooled, imagine how many times they were fooled and fraud wasn't caught" (Augenbraum 2008)? And imagine how many editors of "ordinary" journals are tricked every year. After all, they are not staffed to the degree of *Nature* and *Science*. In addition, I believe many editors are not vigilantly screening against FFP. In this chapter, we will examine

Research Ethics for Scientists: A Companion for Students, First Edition. C. Neal Stewart Jr.

a few real case examples of research misconduct with an eye towards prevention, so make sure to judge yourself throughout the chapter. Of special interest is the creation or alteration of illustrations or figures in papers and grant proposals using image manipulation software. Do's and don'ts will be emphasised when creating scientific images to avoid the appearance of fabrication and falsification.

Why cheat?

Successful scientists have the knack of coming up with great ideas. Great ideas are the foundation of scientific inquiry, but just having a great idea does not necessarily equal great science. Great science is defined as a great idea coupled with rapid and thorough discovery leading to diligent publication of the work that is regarded by and heuristic to other scientists. That is, the formula is great idea + rapid execution + sound data = great publication. Stating the formula is easier than realising that great publication. Finding the big idea is tough enough. This is complicated by needing to be in the right place at the right time with the right people. Then the right people must execute science in such a way to get both the rapid collection of thorough and sound data. The summation of these leading to a breakthrough discovery and stellar paper is a rare occurrence indeed! I believe that the first part of the equation, a great idea, is often a much easier bird to catch than the second part – rapid and sound data. After all, I have come up with great ideas (or at least they seem great at the time) in the shower or at my desk, or walking through the woods. But to make science really happen is much more difficult and takes a lot of thought and work to acquire honest data.

First, you have to find the funds to do the research, and then you have to recruit the right people with the right skills to make the science happen. After these, the experiments have to "work" with no technical problems. And finally the data must be positive, i.e., demonstrate that the idea was sound and typically support (fail to reject, in science lingo) hypotheses tested. This concept of "positive data" is contentious in as much as "negative data" are typically not readily publishable in many fields of science. For example, let's say that I think that a certain technique using a novel chemical addition can lead to the facile genetic engineering of an animal cell. I can test my hypothesis that adding the chemical will increase the transformation frequency. If it does, then my paper will be publishable, but if not (negative data), then almost no editors or peer reviewers will be very excited to publish the paper. Why? Because they say it adds no new knowledge. In a way, that notion is correct. In another way, it is incorrect, because publishing an ineffective method and the sound data could save someone else much time and effort when they come up with the same "great" idea. The world of science rewards great ideas and speedy and elegant experiments showing those ideas have merit.

Speed is of the essence. So, scientists with great ideas and big labs have therefore solved a couple of crucial problems in executing experiments en route to the great paper. They probably already have some funds they can allot to small, but potentially impactful projects, and they probably also have accomplished lab staff (students, postdocs, etc.) who might be willing to participate in a potentially great discovery. Then the trick is to design the right experiments and hope that they work. When they do, it is magical and such experiences can define a scientist's career. The classic example of a great idea and speed is chronicled in James Watson's *The Double Helix* (1968) in which he tells the story of how he and Francis Crick pursued the elucidation of the structure of DNA, with special emphasis on their race against other scientists to make the discovery. Indeed, they used their and other people's data to synthesise the model that would prove to be the correct representation of reality and were rewarded with a Nobel Prize a few years later. Their competitors didn't win the Nobel. While there are arguments whether their process and the subsequent award were entirely ethical, the data were sound.

I believe that cheating in the form of data fabrication is often functionally envisaged as a way to replace rapid and thorough experimentation. There are two types of instances when this seems to occur, at least in the high profile cases. First, and curiously, many experiments do go well – they demonstrate that the idea is a great one and that the scientist does have insight into a problem. As a sidebar, it is important to note that cheaters are typically not incompetent scientists. But instances occur where key experiments fail to yield the desired data. Therefore, there is the temptation to fill in the gap with fabricated data. I believe this is probably the most common kind of cheating. The scientist feels that the work needs to be published, and published rapidly, in the journal with highest impact. Therefore, the argument goes that the means justify the ends. When caught cheating, it is not uncommon for a scientist to emphasise all that was right with a study and to apologise while sometimes attempting to justify ill-gained shortcuts that they believe got the "true" result published earlier than if they had plodded along with subsequent experiments. The second type of cheating is more comprehensive data fabrication. This "Full Monty" of cheating seems to be less common, but it does occur. And when high-profile researchers commit Full Monty data fabrication they are certain to be discovered in their entire naked glory. Perhaps the most infamous case is Jan Hendrick Schön.

Before we get to the case of Dr. Schön, let's take a look at an example of what scientific journals expect in submitted papers. To paraphrase Elesevier's guidelines (www.elsevier.com) of what they believe constitutes honest scholarship, papers should:

1. Be the author's own work and not previously published elsewhere.
2. Reflect the author's own research and scholarship. The paper should be an honest representation of the authors' research.
3. Credit all meaningful contributions of people who participated in the study and paper including co-authors and collaborators.
4. Not be simultaneously submitted to more than one journal.
5. Cite all appropriate existing research, which places the paper in its proper context.

Judge yourself

✓ How do you feel about lying? About liars?
✓ Have you cheated in anything? What was the reason?
✓ Do you trust cheaters? Do you want to work with cheaters?
✓ What is an acceptable amount of fabrication in science?

The case of Jan Hendrick Schön, "Plastic Fantastic"

Scientists share a common plight with all humanity: the temptation to cheat. In 2009, in the midst of writing this book, the floodgates of news stories were open about the sexual infidelity of lawmakers across the US. Amongst the lesser-known cases was Tennessee state senator Paul Stanley, who resigned from his office after reports were made about his affair with a 22-year-old intern. He admitted that it occurred. The reason I mention it here is to showcase the commentary on the affair from a Tennessee state representative Richard Floyd. He said, "You just cannot place yourself in the position to be tempted. Opportunity is the problem, and you get to thinking that the rules don't apply to you." (Associated Press 2009). Or in an even less academic example, Charlie Sheen's February 2011 infamous interview quote where he claims he has stopped pretending that he's not a "bitchin' total rock star from Mars."

The book *Plastic Fantastic* (Reich 2009) is a remarkable account of the Schön case. By all accounts, Jan Hendrick Schön was a hard-working wunderkind, taking a PhD in 1997 in physics from a reputable German University and then moving straightaway to Lucent Technologies' Bell Labs in New Jersey. There at Bell Labs he formed numerous collaborations and worked in the area of experimental physics and materials. The breakthrough research that was apparently realised was in transforming non-conducting organic-based molecules into semiconductors. He published an amazing string of papers on the subject, which contained numerous co-authors during a few years time – at an amazing rate of a major paper every two weeks! His publications in *Science* and *Nature* were heralded as significant breakthroughs that could revolutionise the electronics industry. Schön had become the rock star of science, winning many accolades and major scientific awards for his achievements until the sudden collapse in 2002. Leading up to this collapse were allegations that the results he published were anomalous and not replicable. Closer examination of the papers showed that Schön used the same plots repeatedly in different papers. When asked for his data in lab notebooks, Schön could not produce them. He claimed that because of computer problems, the data were not to be found there either. An independent investigative panel chaired in 2002 found him guilty of 16 instances of research misconduct and he was fired. Indeed, Schön had used the same set of data on several major peer-reviewed journal articles and indeed much of the data were simply fabricated. Some graphs were made using mathematical equations instead of real data. His co-authors were exonerated from any wrongdoing.

The case of Woo-Suk Hwang: dog cloner, data fabricator

Woo-Suk Hwang took a doctorate of veterinary medicine from Seoul National University and then later rejoined the same institution in South Korea as a professor. His area of expertise was reproductive biology and he became known in the 1990s and 2000s for his high-profile work in cloning human cells and mammals, namely cattle and the dog "Snuppy." However, the work that received the most attention, and for good cause – it seemed as if it could revolutionise medicine – was his research on stem cells. Landmark papers were published in *Science* in 2004 and 2005 announcing the first cloned embryonic human stem cells. The first paper indicated that his group had cloned a human embryo and produced stem cells from it after 242 failed attempts (Cyranosaki 2006). The second paper, in 2005, claimed that 11 independent stem cell lines were made from just 185 donated eggs – a much higher efficiency. More importantly, the stem cell lines were supposedly genetically identical, i.e., matched, to different patients. The implications of the research to individualised medicine were immense. It was now feasible to imagine the day where people could have stem cell lines that were custom-made from each person to heal all sorts of diseases using one's own body. In June 2005, when the *Science* paper was published, there was tremendous excitement and extreme Korean national pride in the achievement.

Things went rapidly downhill (Anonymous 2005). In 2004, questions were raised about how and where Hwang acquired so many human eggs. In 2005, an anonymous source contacted a Korean television network with a tip that many of the eggs came from laboratory personnel who were coerced into "donating" for the cause. This would have meant unnecessary surgery for the women donors, and would be a violation of bioethical standards. The television show, which aired in 2005, had a polarising effect. On the one hand, it mobilised Hwang's supporters buoyed by national pride; Hwang had a plethora of fans. Some women even pledged egg donations. On the other hand, it caused many people to scrutinise this particular research, and research in general, even more closely. No one had replicated the 2004 work, and nor had even had the opportunity to replicate the 2005 research. Another shoe fell when a co-author of the 2005 paper, a US professor, attempted to withdraw as an author because he believed that there was significant research misconduct in the Korean lab. *Science*, however, would not allow his resignation as an author, since he had tacitly agreed to a statement that all authors in the journal acknowledge: essentially that all authors take responsibility for the integrity of the research. "Too late," said *Science*. In December 2005, rumours of data fabrication led to an internal investigation by Seoul National University, and to make a long story short, patients and their supposed cloned stem cells lacked a DNA match, indicating no cloning had happened. Further analysis uncovered image fraud in the 2005 paper in which parts of the same image of two cell lines are used more than once to represent the other nine cell lines.

The month of December 2005 was a very busy one for Hwang, his lab, collaborators and his university. The inquiry was open and then closed quickly, with the damning finding leading to the retractions of both 2004 and 2005 *Science* papers the following month. Hwang was treated in the hospital for stress-related ailments. I doubt he had a happy Christmas. Finally, December 29, 2005, Hwang attempted to resign from his position at Seoul National University. The university did not accept his resignation and instead suspended then fired him in 2006. Hwang was finally tried and convicted for fraud in a criminal court in 2009 and was given a two-year suspended prison sentence.

There has been much speculation about why and how this misconduct happened. Could it be that Hwang had put too much of himself into his research? After all, he was a self-avowed workaholic, arriving at the university at 6:00 am and not departing until midnight on most days. Could it have been that his self- and government imposed expectations led to data fabrication? His animal cloning work has been shown to be genuine. Could it have been that he over-extended his research into an area that demanded very rapid results that were not feasible to produce? Was he simply a victim of his own unrealistic expectations? After all, there were millions and millions of dollars riding on the outcome of the research. I suspect that all of these factors played a role in Hwang's downfall and serve as warnings to all of us in research.

Judge yourself

✓ What are your feelings about the Schön and Hwang cases?
✓ How do you feel about Schön and Hwang as people?
✓ Do you view any of the pressures they felt as legitimate justification for data fabrication?
✓ Could you envision yourself being tempted by any of the circumstances or personality traits common with these researchers?
✓ Did they receive appropriate sanctions?

Detection of image and data misrepresentation

Falsified or fabricated data and images can be detected in various venues. For example, grant proposal and manuscript reviewers might see potential issues in submitted documents and report them to grant and editorial authorities, who then follow-up. Before we explore the mechanisms and tools for detection, let's backtrack to formally define fraud. Fraud is the purposeful attempt to deceive.

Most scientists seem to be uncomfortable with the term "fraud" in that it does imply intent. Fabrication, falsification and plagiarism (FFP) is the context that the US government's Office of Research Integrity (ORI) defines misconduct. They are more concerned with legality than ethics per se (Shamoo and Resnik 2003). Data can be made up or manipulated in such a way that is designed to draw interpretations or conclusions that are not true. There are many publications that go into great depth detailing nuances of misconduct. Two publications of note are a 2009 National Academy of Sciences report entitled "Ensuring the Integrity, Accessibility, and Stewardship of Research Data in the Digital Age" and "What's in a picture? The temptation of image manipulation," an article written by two editors of cell science journals (Rossner and Yamada 2004). For our purposes here, my simple objective is to briefly examine what is allowable and not allowable for the digital manipulation of images, and then to understand how it might be detected.

First of all, I readily admit I'm a dinosaur when it comes to image manipulation. When I was a graduate student (late 1980s to early 1990s), we ran gels and took Polaroid photos of them. The developed picture of the gel or Southern blot was then pasted to a piece of paper that contained the figure labels. Then, another picture was taken – typically using a 35 mm camera, again, with film. Rolls of film were sent out for developing and the result was physical slides or prints. Typically, multiple exposures were made in taking the photographs to give the best chance of having pretty pictures. At the time, all this photography for image capture seemed to me to be an art that I wished to avoid, hence the big numbers for repetition. Laboratories with good funding would outsource some of the art to studios that resided at a university. As a result, these professionally-rendered pictures were discernibly prettier than mine. Images of microscope slides or organisms were treated similarly. Sometimes pictures were drawn freehand and a photo was taken. What you saw was what you had. I suppose there would have been a way to doctor them to show something different, but it never occurred to me to do so. Essentially, all the images and the entire manuscript or grant proposal was submitted on paper and not electronically as they all are nowadays. Multiple paper photocopies in fact, with glossy figures were submitted to journals or granting agencies where the copies were subsequently snail-mailed out to reviewers. I do recall wishing that I could make my data-based images look better. I felt like the artistic scientist had an advantage over me, the artistically-challenged scientist. The background hybridisation on a Southern blot and the ill-contrasted microscope picture from my own hands were sometimes not very pretty. When published, it might have sent the message that my science was a bit sloppier than ideal or that I didn't care about appearance as much as I should. The only real alternatives to submitting the not-so-great pictures were to repeat an experiment and hope for cleaner results and better pictures or to not include certain data and hope that the story would be sufficiently told. On more than one occasion a reviewer pointed out that perhaps an experiment needed to be repeated to obtain a figure that would be of "publication quality." I hated hearing that since it meant repeating an experiment that to me would only result in a prettier picture and not better data.

As I think about it, I hardly see any ugly data pictures in journals these days. Does that mean the data are all cleaner and experiments better, or is something else going on? Today, the temptation to manipulate images using Photoshop or other image software must be much greater than in the good old days of physical pictures on paper. A researcher can now "clean-up," crop, merge, highlight, label, and outright change images to suit any whimsy. What is allowable and disallowable? What are the best practices?

The overriding principle for images is that no action should be done to an image that could lead the reader to an interpretation or conclusion that the raw picture or data do not warrant. Obviously, features should not be added or deleted from images. If brightness or contrast changes must be used, they should be used judiciously, and then to an entire image. Non-linear correction, such as gamma correction, should probably never be used, and if done, it should be disclosed in the figure legend (Rossner and Yamada 2004). Essentially all image manipulation should be disclosed in the text of manuscripts. Yet, I don't see much disclosure of this sort in publications.

How can figure fraud be detected? It might be difficult for the average reader or reviewer to see problems in manipulated figures, but editors and office staff who handle a lot of manuscripts acquire an eye for data and figure manipulation. Some people in my lab have found duplicated images from previously published work in submitted manuscripts without even looking for them. In one instance we were really interested in a certain new field for us, and so we jumped at the chance to perform a peer-review of a tissue culture paper using one of our new favourite plant species. In this submitted paper, the authors cited one of their previous papers. In the previous paper, they had used one set of constituents in their tissue culture media. In the new paper, they used another set. We were happy to have discovered their previous paper, since we weren't aware of it before we performed the peer review of their new paper – they cited the old paper. Lo and behold, we noticed that there were duplicate images in the two papers. Publishing the same images in two different papers is also not allowed, because data and displays are assumed to be novel and unpublished at the time they are submitted. They are assumed to represent different experiments. After all, if they are identical, why publish the second paper? This requirement of novelty in publishing is especially true of peer-reviewed research papers. It is also often a copyright violation to reprint pictures without the permission from the copyright holder, typically the journal or authors. In this case, however, the authors were inferring that the tissues pictured in the first figure were those described in the second paper. Of course, the same pictures appearing in the first paper claimed the same thing about the first study. Since the pictures were identical, it was impossible for at least one of their claims to be true. So, as peer-reviewers we alerted the journal editor of the misconduct. Once contacted, the authors might have replied to the editor, "This was just a mistake. We intended on using another figure and inadvertently included one that was already published. We have the correct figure – here it is." While this explanation might indeed be factual, these authors' credibility is now quite low in our eyes. We'll naturally be on the lookout for other papers by them and, naturally, we'll scrutinise their science for FFP. As reviewers, we have seen this type of misconduct now twice, and found it when we were not even looking for research misconduct. We were simply gathering information for our own science and happenstance led us to the duplicate figures. Indeed, for the Woo-Suk Hwang case, the editor of *Science* did not believe that it was likely that reviewers could have seen the problems in the submitted manuscript (Couzin 2006b). " 'Peer review cannot detect [fraud] if it is artfully done.' " The paper was examined by nine reviewers – six or seven more than is typical for most journals. Plus the editors read and reviewed the paper. Editors and reviewers are typically not on the lookout for misconduct. Only when a procedure can't be repeated or the science itself seems faulty do scientists typically begin looking for misconduct. In many ways, scientists are a trusting lot. They generally expect that honesty is a shared value among fellow scientists. Editors should probably not be so trusting, especially when image misrepresentation is so easily carried out. Indeed, some editorial offices are examining figures more carefully and subjecting them to image analysis for the purpose of detecting manipulated data. Image analysis can detect if a composite figure has been produced, whether elements have been added or deleted, and whether a figure is "too good to be true," but it cannot detect figure duplication of the sort that we found out during peer review of submitted manuscripts – at least not yet.

With data integrity at stake, it is important that researchers are honest with data. As Rossner and Yamada (2004) point out "creating a result is worse than making weak data look better,"... but these are both forms of misconduct. Accurate representations of study are clearly expected by readers and the science community. My lab does not own a copy of Photoshop and I have no intention of purchasing it now. Why tempt temptation?

Judge yourself

- ✓ Would you be tempted to overly manipulate an image to make it prettier or to avoid repeating an experiment?
- ✓ Do you enjoy manipulating images?

Neither here nor there – the curious case of Homme Hellinga

As stated in the first part of this chapter, papers are retracted for various reasons that have absolutely nothing to do with misconduct. Shortcuts in experiments, misinterpreting data and simple mistakes happen all the time in science. Since science is largely self-correcting through peer review, careful analysis of the literature and data, and replication of experiments, I believe that many papers containing erroneous data are identified. The ones with "fatal" mistakes are either not published or retracted after other scientists and journal editors point out fatal flaws.

Homme Hellinga is the James B. Duke Professor of Biochemistry at Duke University Medical Center. His interest is in protein design and he uses experimental and *in silico* techniques to unravel protein structural relationships to function. In 2008 Hellinga retracted papers in *Science* and the *Journal of Molecular Biology* (Hayden 2008). The *Science* paper was accepted by *Science* less than 1.5 months after it was submitted in 2004. According to Hayden (2008), the paper only included Hellinga's graduate student's "best data" that demonstrated that a non-enzymatic protein had been converted into an enzyme – a triose phosphate isomerise (TIM) that Hellinga and colleagues called a NovoTIM. It was later discovered by a researcher at the SUNY-Buffalo that the so-called engineered NovoTIM was most likely simply an endogenous bacterial TIM. The Buffalo researcher emailed Hellinga in 2007 to inform him that his results could not be confirmed by the Buffalo lab, and also copied the editors of *Science* and the *Journal of Molecular Biology*. Hellinga then accused the graduate student of data fabrication. She was still at Duke, but by then was working in another lab as a postdoc. After meeting with his former graduate student (she finished her PhD with him in 2004), Hellinga forwarded the matter to Duke University, which did an investigation and exonerated the student of any wrongdoing.

While, on the outside, it appears that the matter devolved into a he-said, she-said matter, at the crux of the situation is an example of where science went wrong and was then handled poorly, but apparently there was no FFP. We may never know exactly what went wrong with the science, but it is certain that professional ethics were suboptimal in this case. It seems to me that if Hellinga indeed selected the "best data" to go in the paper, then that was a mistake unless it was completely justified by the science. In this case, it obviously was not. Scientists make decisions constantly about which data to include and which to exclude, and we've all made judgment errors. But accusing an innocent graduate student is perhaps the bigger mistake. As we saw in the previous chapter on mentoring, professors have nearly all the power in the mentor-trainee relationship, especially when it is an advisor-student relationship. Professors ought to wield their power carefully. In 2008, Duke University commenced a Homme Hellinga investigation, and at the time of this writing – some 15 months later – the investigation is not over (Borrell 2010). In this case, Hellinga requested that Duke investigate him.

I don't know Homme Hellinga and I don't know the details of this case beyond what is reported. But I do know that a lack of humility will lead to trouble. Hellinga has been accused of being arrogant; what I refer to as the God Complex. Hayden (2008) referred to a colleague to whom Hellinga asked, "Do you think I'll be more famous than Darwin some day?" Reading his tale causes me to pause and reflect about how I, as the PI and typically the corresponding author on papers, handle students' data. Even more importantly, I think about interpersonal integrity, including the consideration of my trainees' feelings and their career futures.

It doesn't matter to me how brilliant a professor is; there are some things that are intrinsically worth more than a paper in *Science*. One potential problem of publishing any paper, much less one in an extremely high impact journal, is that others base their livelihoods on the findings. Herein lies the problem. If a paper is sound, then follow-on science by the same or other groups contributes greatly to society. But if a paper is rushed through the submission, reviewing, and/or contains faulty, or, in the worst case, fabricated data, science is not served and it can crush a graduate student's or postdoc's early career; maybe knocking them out of science.

Judge yourself

✓ How would you feel to find your PI in trouble – either rightly or wrongly accused?
✓ How important is it for you to trust others' research findings?
✓ How much due diligence is appropriate when reviewing a submitted paper or simply using the information published in a peer-reviewed paper? How can we know that the information is true? How much sleuthing do you want to do?

Lessons learnt

Why do up and coming science superstars like Jan Hendrick Schön and Woo-Suk Hwang cheat? Cheat at the highest levels? By all accounts both of these researchers were competent and productive prior to cheating. Did they believe that the means justify the ends? That is, perhaps what they believe to be truth really is true, and the pressure to establish the facts using science is too slow for their satisfaction. Perhaps they believe the rules of science apply to "ordinary scientists" and not themselves. Perhaps they became addicted to *Science* and *Nature* publications. Whatever the case may be, and I think there are multiple causes, research misconduct is infuriating to both the straight-players in science and the public. In all of the cases we examined, sloppy or fraudulent science cost innocent researchers much time and effort in trying to replicate the unreplicable. These innocents are typically graduate students and postdocs whose time was wasted and careers delayed. And what about innocent parties within the misbehaving PI's lab? They are now often deemed guilty by association. After all, who would even trust a letter of recommendation from Woo-Suk Hwang? The price of research misconduct is hefty and its turbulence runs deep with collateral damage.

So, how are we supposed to respond if we detect research misconduct? I gave one brief example in the case of reviewing submitted papers for journals – simply alerting the editor of a probable misconduct. We'll look into both the more common forms and few less simple (and more painful) examples of "whistleblowing" in the next chapter.

Judge yourself *redux*

✓ What are your feelings about the Schön and Hwang cases?
✓ How do you feel about Schön and Hwang as people?
✓ Do you view any of the pressures they felt as legitimate justification for data fabrication?
✓ Could you envision yourself being tempted by any of the circumstances or personality traits common with these researchers?
✓ Did they receive appropriate sanctions?

I feel saddened by these two high-profile cases. As people and scientists, they were obviously competent and very industrious. Unfortunately they made a series of very bad choices. Everyone has pressure, but these two individuals, especially Hwang, placed a tremendous amount of pressure on himself, which likely led to the circumstances setting up data fabrication and other ethical issues. The lesson for me is to not allow such pressures to be created on myself. Of course there can be temptation, and of course we sometimes do things under duress that we normally would not do. I'm not making excuses for them (or me, should I have ethical blunders), but I can understand how these things could happen. I think their sanctions were appropriate – certainly not too harsh for the magnitude of the misconduct.

Judge yourself *redux*

✓ Would you be tempted to overly manipulate an image to make it prettier or to avoid repeating an experiment?
✓ Do you enjoy manipulating images?

Of course I'm tempted to make the data look better. We all are. No one likes to repeat experiments to get a nicer image. Earlier in my career, I loved doing image manipulation for fun. On my webpage, I once put the picture of one of the graduate students "behind bars" because it was taking him a long time to finish his experiments. I figured this would help him speed up. It didn't.

Judge yourself *redux*

✓ How would you feel to find your PI in trouble – either rightly or wrongly accused?
✓ How important is it for you to trust others' research findings?
✓ How much due diligence is appropriate when reviewing a submitted paper or simply using the information published in a peer-reviewed paper? How can we know that the information is true? How much sleuthing do you want to do?

I know that I would feel bad for my boss as a person. I would be as supportive as possible. But at the same time, what's right is right and in science we have to trust research findings to be true and sound. I can spot sloppy science. I think most scientists can. If a paper is too sloppy and there are obvious issues with quality, I tend not to hold the findings in high regard. But I remember that it took me several years before I was able to develop a "standard." I tend to not like to spend much, if any, time playing detective on bad papers.

Summary

Falsification is probably the most heinous of research misconduct and can cause damage to the perpetrator's career, but also to those of trainees and other scientists who trust research findings are true. Not to be confused with honest error, falsification is a sanctionable offense. Image manipulation is becoming especially problematic given available software. Be careful to declare all image manipulations. Don't allow too much pressure to put on you that could lead to a greater temptation to cheat.

Chapter 7

Research Misconduct: Falsification and Whistleblowing

ABOUT THIS CHAPTER

- One of the most difficult decisions a scientist makes is when and how to report a case of research misconduct.
- The decision has important consequences for the reporter and reportee as well as other people in the lab and university.
- Reporting misconduct discovered as a reviewer is easier than blowing the whistle on a colleague, but this should not be done casually.
- Science integrity is our corporate responsibility, but there are many procedural considerations to be weighed up when contemplating blowing the whistle.

Integrity is tested when misdeeds are observed and a scientist is left with a difficult decision: report misconduct, or not? How is it done? What about the special instance in which it is believed that a student's colleague or mentor is making up data? What are the consequences? Whistleblowing is not for the faint of heart, since there are many documented examples of retribution and unintended negative consequences that do not favour the whistleblower. Indeed, great courage is required by responsible scientists to face unpleasant facts and do the right thing for the sake of science integrity. This chapter is perhaps the most important one in this book inasmuch as it, in many ways, defines professional ethics in research science. The chapter integrates mentorship, responsible conduct in research, grantsmanship, research pressure, with the responsibility we all share as citizens in science.

Whistleblowing takes its meaning from reporting wrongdoing, e.g., a crime in society or breaking the rules in a game, say, when a policeman blows the whistle on a criminal or a referee whistles a foul in sports. In all cases, the perpetrator is not very happy with the whistle being blown and many other people are often dismayed as well (e.g., think about your reaction when your favourite football team is called for a penalty). Almost nobody seemingly appreciates the whistleblower. If the person being reported is popular, personable, and had apparently lived an

Research Ethics for Scientists: A Companion for Students, First Edition. C. Neal Stewart Jr.

exemplary life in the past, there is a tendency to judge that the whistleblower is simply wrong or vengeful. It is true that being wrongly accused of wrongdoing is hurtful towards one's reputation and career. Whistleblowing is not something to be done without great consideration. The possibility of errant whistleblowing is also threatening and a factor in non-collegiality. After all, if the whistle can be blown on my colleague, then perhaps I'm the next target. Reporting FFP should always be done after very careful contemplation and with due diligence.

We will examine one of the most publicised true cases of whistleblowing in recent years and its ramifications on the people involved. In this case a University of Wisconsin professor was reported for research misconduct by her graduate students. The case is true and documented in at least two reports (Couzin 2006c; Allen 2008). It does not have a happy ending for any of the involved parties. To a person, their studies and research projects were severely altered or abandoned, and in most cases, they left their university studies under duress.

Understanding the typical structure for maintaining research integrity is important when thinking about blowing the whistle. In universities, typically the chief research officer is ultimately responsible for investigating accusations of research misconduct. In companies and research institutes, a parallel position would exist for that duty. Typically, universities do not have standing committees or panels to investigate and judge cases of alleged faculty misconduct, which contrasts to the situation with students. Most universities do have undergraduate and graduate honour systems. Why the difference? It is simply a matter of frequency of cases. It is rare that cases against faculty crop up. We will see some reasons why later in this chapter.

What is the path of whistleblowing? If students or colleagues report, typically, a bottom-up approach is used. Cases get reported to department heads, who might informally investigate, before notifying a dean who might also informally investigate prior to notifying the chief research officer. The research officer usually is the person responsible for the administration of the university's contracts and grants office and research programmes. This person could get notified of a possible infringement by a grants program manager or editor; these are two of the other key people in the horizontal whistleblowing chain.

Before we examine this most painful case of internal whistleblowing, external whistleblowing will be discussed. Scientists are responsible for maintaining truthful reporting of data; their own and others. It is our corporate duty as scientists to tell the truth. Indeed, not reporting misconduct is also unethical (Chalmers 1990).

External whistleblowing is much more common than internal whistleblowing, i.e., reporting a professor within a shared institution is rare. As an external whistle-blower, I have notified editors and grant program managers of potential problems I observe in documents so they can follow their procedures. But I have never reported anyone from my own university. There are three obvious reasons for this. First of all, I have more opportunity to see problems with papers or grant proposals in the outside and larger universe of science than misconduct perpetrated by my own colleagues. Second, the author of a bad paper or grant proposal outside my institution might not be known by me. A colleague is more likely to be a friend. Friends typically don't blow the whistle on their friends. In the latter case, I might be more willing to confront a friend informally in hopes of remedying a bad situation; I wouldn't likely take that option for an unknown person in another time zone. The third reason has to do with the nature of the editor/grant officer-reviewer relationship. The reviewer is basically charged with finding errors in submitted papers and grant proposals. That is, reviewers are invited to dig around to unearth issues. While most reviewers don't specifically look for FFP (I don't), sometimes they find it anyway and, then, should report it along with the normal errors that all reviewers denote while reviewing. As we'll see in a later chapter, all peer-reviewing done for papers and grant proposals is typically performed anonymously; so ethics would not even allow me to informally intervene in a paper submission – even if I wanted to. For external whistleblowing – a paper or a grant, the whistleblower is nearly always completely protected from any negative ramifications of reporting misconduct; i.e., whistleblower fallout. The editor or grant officer may or may not constructively do anything with the reported information, but at least the whistleblower is not penalised. The structure of reviewing that features anonymity does favour objectivity and honesty while protecting the reviewer. Internal whistleblowing is not so straightforward; it is definitely not painless.

A "can of worms" indeed: the case of Elizabeth "Betsy" Goodwin

The profile of a successful professor

> "Goodwin was one of the rising stars of the UW's genetics department. "She was an extremely good citizen," says Phil Anderson, another *C. elegans* expert and a professor of genetics [at the University of Wisconsin-Madison]. 'She did more than her fair share of committee work, and she was very involved socially. She entertained prospective graduate students and helped recruit new faculty. She brought a genuine sense of joy to working here' " (Allen 2008).

Dr. Elizabeth Goodwin had received her PhD from Brandeis University and completed a postdoctoral stint at the University of Wisconsin. She had published over 50 papers, some of which in very high-profile journals, had been successful in grant-getting and had a vibrant lab of six PhD students and a technician. Using every available indicator of success, Goodwin passed with flying colors. In addition, her group also had fun together outside of the laboratory walls. They enjoyed parties and horseback riding adventures among other normal extra-science functions that bring lab "families" together. Professor Goodwin was a hands-on mentor, communicative and called regular lab meetings. The scientists were united in working on a suite of challenging problems at the cutting edge of science.

Her lab was one of the few looking at sex-determination in the model nematode "the worm" *Caenorhabditis elegans*. Every person in her lab actively studied this organism, which is a very powerful developmental biology model. As it turned out, Goodwin hypothesised that one particular gene would encode a small RNA, which she thought was crucial for sex determination. The gene had been linked to the male/female trait in *C. elegans*, but the small RNA hypothesis was novel. As we'll see, her dogged determination to follow this line of research might have led to some of her ensuing problems. Small RNAs were (and still are) a hot topic since many control gene regulation. Another pertinent fact is that five of her six PhD students had been working in her lab a number of years, but with little data to show for it. So in many ways, this case is a "perfect storm" of dynamic researcher, a hot research topic, the need for continued funding – all adding up to pressure – but with little recent success to show for it.

A Goodwin gone bad

Our story begins in the autumn of 2005 (Couzin 2006c). Chantal Ly was a frustrated student in the seventh year of her PhD program. She was unable to replicate certain experimental results produced earlier by the Goodwin lab. As an aside, I've been there with my own advanced students. You can hear their clock ticking – the degree that should have taken far less time to complete is stymied because experiments are just not working out the way envisaged. It is not an easy situation. Dr. Goodwin then does something with which I've had success with my own trainees – she gives Ly part of a grant proposal to spur an interest in pursuing something else that might work, which could then substitute for the unfruitful PhD research. We all want students to experience success. Goodwin posits the proposed research as experiments she'd intended for another PhD student, Garett Padilla. Goodwin thought, however, that there would be room for both students on the grant, should NIH choose to fund it. After the proposal was submitted Ly noticed that there is a figure in Goodwin's proposal labelled as unpublished data that was, in reality already published by Goodwin and co-authors the previous year. That part is not ok – the data in the proposal should have referred back to the published paper. It gets worse. The protein of interest in the published paper was different from the one indicated in the proposal. Ly was so troubled that she shared the portion of the proposal with Padilla, who was also her officemate, to ask his opinion about the possible falsification. As they looked further, they noticed more irregularities in the grant proposal. It had appeared that Dr. Goodwin had falsified data in more than one instance. What would be the student's next step? One certainty was that disillusionment rapidly set in as the students learned that all was not perfect in paradise.

The grant treadmill

Grants are hard to win. Typically for any funding panel as few as 10% of proposals submitted result in awarded grants. Since it is a competition, professors must "sell" proposals to boost their chances of winning. There is a temptation to approach the edge of truth and sometimes go beyond it. This practice breeds skepticism in reviewers who must cut through any potential smoke and mirrors with a sharp scalpel. Proposal reviewers want to see preliminary data that convince them that an idea and approach are sound and merit funding more than other proposals they are reviewing. It is a competition after all.

In the Goodwin case, it appears that the preliminary data went over the line of acceptability and into the world of misconduct. But why would Dr. Goodwin feel as if she needed to cross the line? In two words: competition and doubt. As scientists, we all know that our proposals are competing against other proposals, therefore there is a strong motivation to push the envelope of integrity. It is an unfortunate part of the system of science, but no one has thought of a sustainable alternative to competition to support large amounts of the best research. The assumption is that the best science (best proposals) typically receive the funding. Doubting your proposal's ability to compete is an ever present demon since you don't know anything at all about the competing proposals. You wonder, who's submitting what? Who's on their team? How will the reviewers compare my proposal with the competition? This dynamic situation creates pressure for researchers. In a research university, graduate students, postdocs, and other personnel, supplies, equipment, etc. are largely supported solely from grants. There is definite motivation to write winning proposals. The alternative is a starving lab; no mentors want their researchers to starve. In some areas of science the pressure for funding is greater than others; the more expensive the science, the more pressure for funding. Small amounts of funding are needed to grow and weigh plants, for example, and therefore the responsibility and stress to compete at this level is not as stressful as that needed for sustained biomedical research.

Judge yourself

✓ How do you handle pressure and competition? Can you stay cool or does it keep you awake at night?
✓ Do you enjoy "selling" and being the person in charge?
✓ Do you have integrity to play by the rules?

The plot thickens

Padilla started keeping a log of the case (spoiler alert: he later goes to law school) and then consulted with another UW faculty member, a person in another department with whom there were personal connections. He also consulted with a former Goodwin postdoc who was, by then, working at a company. They both encouraged Padilla to talk with Goodwin about the matter, which he did – on Halloween [cue the scary music]. According to Padilla (Couzin 2006c), the meeting did not go very well. She denied falsifying data, but admitted that mistakes were made. At the end of the meeting Padilla did not feel as if any satisfying resolution had ensued.

After this meeting, all seven lab members were invited to a meeting to discuss the alleged falsification that was held in a building different than that housing their lab for discretion to discuss their plan of action. Choice 1, having a discussion with the UW administration, was dismissed as being too risky. They wondered what would the administration do to Dr. Goodwin? How would it all be handled? Choice 2, sending Padilla back to speak with Goodwin, seemed like a more prudent path. After all, perhaps Padilla, the lawyer-to-be, could discover the real underlying problem and solve it. Maybe she could retract the grant proposal. Indeed, the second Padilla-Goodwin meeting seemed to go better. She promised to email her NIH contact and copy Padilla (which she did) explaining the problems in the proposal. She asked for Padilla's forgiveness. She said there would be little chance the proposal would be funded. No harm, no foul, thought Padilla.

Judge yourself

✓ How would you feel to be in a situation similar to Ly and Padilla?
✓ Are you, by nature, confrontational or do you avoid confrontation?

But wait, there is more

Couzin (2006c) tells of a third student, Mary Allen, who was not convinced the situation was resolved sufficiently for her conscience to allow her to remain in the lab. Even though she was in her fourth year of a PhD program, she was willing to change labs to remove herself from the problematic situation and mentor. After explaining her wishes to Dr. Goodwin, additional reasons and explanations were forthcoming from Goodwin as to what had occurred in the alleged misconduct. The professor claimed to have received unlabeled images from a labmate and had simply mixed up the files. Goodwin admitted to making mistakes, perhaps suboptimal judgments, but she claimed she had not falsified data. After this meeting, the students continued to confab and as a result, grew more and more worried about the future. The proposal to report Goodwin to the administration was forwarded again. Instead of dismissing this idea outright, they decided that they would not visit with administrators unless it was unanimously decided by the lab to proceed. They were also careful not to discuss the matter with others outside the group, except the two people they'd already brought into their inner circle as advisors. As November turned to December, they finally made the decision to report Dr. Goodwin to their department head. The students reasoned that, even though it might go bad for them, Dr. Goodwin might place future graduate students in the same predicament if they did nothing. We can see that by their actions they were obviously convinced that she was guilty of research misconduct. Their decision was a brave one.

Not a great spring semester for the professor

The department head referred the matter to two deans, who informally investigated to see if there was just cause for further investigation, which is the typical path and action followed by university administrators. In a meeting that included the department head, Goodwin and the lab, Goodwin claimed that she "was juggling too many commitments at once." This, certainly, is a common predicament in which faculty members can find themselves. A faculty member has tremendous pressure to find and keep funding, publish papers, mentor students and be a good citizen of the university and to science. She claimed the bad figures were just placeholders for good figures she had intended to substitute in the proposal. I wonder about the figure legends? It seemed that the explanations did not jibe with reality. The informal investigation turned into a formal investigation with the university appointing three professors charged to uncover the truth. Past Goodwin grant proposals that were funded were also inspected. It turned out that funds from all three grants were sent back to the NIH and USDA.

By the end of it all, Elizabeth Goodwin resigned March 1, 2006, not even halfway through the spring semester. UW officials were obliged to continue the investigation to assure that they had uncovered all the problems, a task made more difficult by Dr. Goodwin's absence. Another notable fact is that the US Federal Bureau of Investigation (FBI) became involved in investigation (Winter 2010).

The fallout

Even though their professor was largely absent all semester, even before her resignation, the students attempted to carry out their research and live their lab-life as usual. They did experiments and held lab meetings sans Goodwin. They were assured by the administration that their salaries and stipends would not disappear that semester. And indeed, all of their funding remained intact during the short term. But a lab without a PI is like a symphony orchestra without a maestro – successfully executing Tchaikovsky's Fifth Symphony is not an option without the maestro and neither is high-level science possible without the PI. Except that this musical analogy underestimates the situation. Graduate students cannot complete their degrees without a mentor to sign-off the PhD. Whether the mentor dies, resigns or does not get tenure, there are limited choices for graduate students to complete their research projects and degrees. For example, when I switched universities during the midstream of two graduate students' degree programmes, they both opted to move with me. I'm not sure what their options would have been if they'd decided to stay at my first university. If a professor simply resigns or dies, students must either find another professor to finish them up or they can start over. These choices were the ones that awaited the three graduate students, but there also seemed to have been an element of "contamination." University departments are cozy entities. Few people want to risk contamination of perceived "damaged goods" into their labs. My choice of words here is very stark, but in the worst of situations, many faculty members would own this perception. In the best of outlooks, the system of graduate student education at most universities is simply not set up to handle deviations from the traditional mentor-student relationship. Even then, the relationship can become strained. Few faculty members aspire to enter into a mentor-student relationship that is from the outset odd or strained. The risk-benefit analysis is usually done to arrive at a conservative decision when admitting new people into the research lab.

My second best favourite quote (the first uses a word not typically found in the pages of *Science* – find it for yourself) from the Couzin (2006c) article was about how many of Goodwin's faculty colleagues felt about the situation. "Goodwin had had 'to fake something because her students couldn't produce enough data.'" I'm sure the faculty rumour mill had been working overtime to attempt to explain the problem away and exonerate Dr. Goodwin. Faculty members are famous for looking out for their own interests and personnel. Human nature demands that they likely didn't want to face the fact that their colleague was possibly trying to prove something that was not scientifically possible and that fact was the real reason for the students' unproductivity. So, given that the above view was rampant, I don't imagine the UW faculty were falling over each other trying to recruit the students. Even if they were, the students would mostly have had to start over with new projects. However, two of the students did stay at UW and had successful "do-overs." The other four PhD students left. Only Mary Allen continued on as a PhD student. Padilla went to law school and Ly is working at a company.

Judge yourself

✓ Did the Goodwin debacle end the way you think it should?
✓ What are the alternative endings to the story?
✓ What would you have done differently if you were in the students' position?

Life is so unfair…to the whistleblower. Is it possible to be a whistleblower without being ejected from the game?

In retrospect, the students mostly did everything "right" with regards to ethics but enjoyed few, if any, direct benefits from being good ethical citizens in science.

They did not take lightly their difficult decision to blow the whistle. They kept, to a large degree, the situation confidential. Finally, after the students decided to report, they reported to the appropriate university administrator; their department chair. For their efforts, their degrees were largely scuttled and the time and effort they had devoted in pursuit of the PhD was wasted. I'm not sure how it could have ended differently, however. Such is the life of the whistleblower. But they all fared better than Goodwin. She was fined $500, ordered to pay a total of $100,000 restitution, and was sentenced to two-years' probation after being found guilty of making a false statement (Winter 2010). I don't know what she is doing currently, but I don't think she is in science.

Science is not alone with regards to unfair treatment of whistleblowers; police departments, schools, politics, and businesses have all seen unfair persecution of people who point out injustices and wrongdoing in the professional world. As Gunsalus (1998) points out, society has a "visceral cultural dislike for tattletales." Indeed, if the whistleblower gets it wrong or is vengeful, an innocent person obtains a sullied reputation or worse. Thus, the motivations and conscience of the teller are important; from the outside, these might not be evident or even knowable. Certainly, believability and trust in the whistleblower are keys, regardless of motivations. This is why the utmost care is needed when assessing such a situation and taking action.

Take a fictitious example of two evil brothers. When one evil brother tells their father about the evil deeds of the other brother, the dad is often dismissive of the event. Yes, he might believe that the perpetrator had done an evil deed, but the whistleblower, also evil, has low credibility as a reporter and the father does not want to be dirtied in the mess. What about the situation when just one of the two brothers is good? When a good brother tattles on the evil brother, experience has shown that the good brother is usually trustworthy and right. In either case, however, the father might prefer for the brothers to come to justice without his involvement. Let's contrast this case of brothers to a case in science, such as the Wisconsin incident. No matter how you slice it, it is impossible to paint a scenario where graduate students can benefit from their major professor's professional demise, especially when the group of students act together to report. It matters little whether the graduate students are viewed as "good" or "evil." Therefore, we can eliminate any motivations of retribution or personal dislike in such a case, since it is certainly to the students' disadvantage to report their professor's misdeeds. Even if their ethical motivations prevail, they still could have been mistaken in their accusations, but again, since they acted in concert with due diligence, this scenario is unlikely either. We are faced with the facts that it is very unlikely that any graduate student or postdoc will unjustly report his or her mentor for anything. I recall a certain graduate student whose mentor had caused him all sorts of personal pain and anguish during the course of his PhD degree program. In a particular moment of brutal honesty (and after a few glasses of wine), the student confessed that he would not hesitate to murder his mentor by strangulation, except that he wouldn't get his PhD if he did it.

A university administrator's opinion

While universities might wish to protect whistleblowers who do the right thing and report serious cases of misconduct, the deck is stacked against the reporters, especially when the reporters are students. Administrators know that. Let's face it. A graduate student, whose role is most likened to a type of intern or apprentice is in a position of great inherent weakness relative to his major professor, especially if the professor holds tenure. While it may seem that professors are also relatively weak compared to administrators, this is not necessarily the case since tenure is powerful in maintaining the *status quo*. One important role of society is to protect the weak. Why should a university society be any different than society at large? As we observed in the case above, the close mentor-trainee relationship of graduate education results in the intertwined destinies of professor and student; especially in the case of failure by the professor. In addition, this relationship must be built upon mutual trust. I can't see any way around this situation in today's graduate education system. Considering these circumstances and others, let's look at Gunsalus (1998) six rules for responsible whistleblowing.

Rule 1 is to consider various explanations about a potential incident, including "especially that you [as the whistleblower] may be wrong." It is critical that a complainant think big and broadly about an incident of misconduct. It is important to understand all the facts and to be able to explain them. None of us is omniscient, and there are typically mitigating circumstances to consider.

Rule 2 is critical: "ask questions, do not make charges." No one likes to be accused of wrongdoing, and in light of Rule 1, that the reporter could be wrong, it is important to exercise the student's inquisitiveness rather than the victim's judgmental side. In the midst of what could eventually be whistleblowing it is important to ask as many questions as possible and to make known that questions are merely being asked. You don't understand but you want to. Listening is more important than talking when asking questions.

Rule 3 is to focus on facts and data. Without these, there is no case and people are more apt to focus on the whistleblower than on the whistleblowee and the potential misconduct. If the data are not available to the reporter, they might or might not be available to others. Evidence is crucial and cannot be overlooked.

Rule 4 is to separate personal and professional concerns. Students seek to be professionals in science, and it is important to conduct investigations into FFP in as factual a way as possible, even when emotions come into play. This is not to say that a complainant must be dispassionate since professional ethics is something that many people ardently protect. It is important to focus on deeds (or misdeeds) rather than the personality of the perpetrator or your own personality.

Rule 5 is that goals should be assessed. A potential whistleblower might wish to simply change a situation informally rather than lodging formal charges; behaviour change could be more important than punitive charges. Of course, in light of the fact that misconduct should never have happened to begin with, one objective of the reporter is to ensure that it should not happen again with this mentor. Here an administrator could reasonably ask a complainant what they want the administrator to do about it. These goals should be elucidated.

Rule 6 might be the most important of all: "seek advice and listen to it." A student should realise that university administrators have seen more problems and have a larger scope of professional insight than has the student. Other people might also give advice: other students, trusted colleagues, relatives in relevant positions; but the confidence concerning the potential reportee needs to be maintained.

In summary, it is important that a potential whistleblowing student learn the facts, ask questions, seek advice, listen to the answers, and be professional.

In light of these rules, Gunsalus (1998), an administrator at a large research university, offers step-by-step procedures for responsible whistleblowers. Again, I will frame these steps with the student or postdoc in mind.

The first step relates to Rule 6: share concerns with a trusted person. Everyone needs a trusted colleague who can lend a non-judgmental ear. This process allows the potential reporter to understand the gravity of his or her position while simultaneously gaining insight and advice. It is a rehearsal of formal complaint too. Again, asking questions rather than levelling charges is a powerful and prudent approach. A large part of trust is confidence, so it is important to be assured that information will remain confidential. It is also important to choose your people to trust very carefully. You would not want to choose someone with a conflict-of-interest or a personal friend of the whistleblowee.

The second step is to listen to the advice of your trusted colleague. The person might agree or disagree. Either way, it is important to really listen and to be objective. Hear what you say through another's perspective – gain their ears if you can. Try to read the situation and your trusted advisor's comments. Take them seriously. If a reporter still holds a strong opinion that an official complaint should be lodged, then the next step must be taken, but carefully.

The third step is "to get a second opinion and take that seriously too." Gonsalus makes the point that even large research universities are small communities. It is easy to "get on a roll" of "confiding" in people so that a gossip chain is initiated. That is the last thing a potential whistleblower needs. But like step 2, listen to the second person's opinion, how the facts are reflected, and focus on asking questions not making complaints at this point. This might be a teller's last real chance to tell the story without negative consequences. Formal investigations are not for the faint hearted and to be sure, there will be negative consequences for the whistleblower.

Step 4 occurs once a decision is made to initiate formal proceedings: seek strength in numbers. As we saw with the Wisconsin case, the entire lab together reported their professor. If there are natural groupings of people who share your concerns, then perhaps they would also share responsibility as complainants. Again, sparking off the rumour mill should be intently guarded against. It works against the process and the whistleblower by unnecessarily harming the person to which complaints might be made. Gonsalus makes the point that if people don't wish to join in a complaint then the primary complainant should assess why not. It might be informative as to how an official might receive the case and should be factored into whether to go to Step 5.

The fifth step is to find the right place and procedure to file a complaint. As a student, a department head is almost always a good starting point. The head can illuminate procedures and again, act as a sounding board. It might be that a potential complainant will be dissuaded from going to Step 6.

Step 6 is officially reporting a concern. Again, this should focus on the facts, a question, or neutral observations. It is best to avoid acting as judge and jury. As much of the evidence as available should be given.

The seventh step is to take notes and ask questions. It is important for whistleblowers to protect themselves by keeping informed and active in the process. Keep a journal of the same quality as you would a lab notebook with dates and numbered pages. Gonsalus points out that it should not be necessary to seek council of an attorney. In addition, never speak to the press about the case!

The last step is to exercise patience. Investigations and procedures will take some time to complete, as they should. Stay involved and be patient. Hope that the FBI doesn't get involved!

I will add one additional step in addition to Gunsalus's excellent list. Any student who decides to go the route of filing a formal complaint should have a support group. After blowing the whistle life will not be easy for anyone. Being a graduate student is already hard. Becoming a graduate student who is perceived as being ungrateful, not loyal, and other unpleasantries (fill in the blank) will likely need all the personal help they can get. And they will also need to look for another mentor, likely to be located far away.

Judge yourself

✓ Do you have a support network? Are there people in your lab or network that you trust to share sensitive information?

✓ Do you know your department chair and dean?
✓ Do you have the kind of relationship with your major professor that enables frank discussions?
✓ Would you ever envisage yourself as a whistleblower? How bad would an event have to be to report it?

Deal with ethical quandaries informally if possible

As we see from the case study, becoming a formal complainant typically results in a one-way path to a place most students and postdocs did not initially chart for themselves. Therefore, my best advice for potential internal whistleblowers is to proactively respond to potential misconduct issues early, quickly, and informally. This procedure also takes courage, but can reflect much better on the informal-than on the formal whistleblower. I will offer a true story, and one from my own lab (but names were changed), to demonstrate how powerful informal intervention can be.

To begin the story, my style is one that values delegation, interaction and sharing ideas, the blame and the glory. Since I started teaching research ethics, I've told my students and staff to ask plenty of questions and to feel free to discuss ethical considerations. Students and postdocs have asked me about various scenarios and situations – to find the ethics in them all. I feel some sense of professional pride in being able to answer in the affirmative or to review how we do things to make governing a situation more ethical. Still, some students are not absolutely comfortable in asking me questions of an ethical nature, especially if they think I might construe questions with confrontation. I understand the potential awkwardness.

"Jill" was a new masters student in my group who had just determined the direction of her research. It focused on the overexpression of a plant gene (P) in a crop plant using a biotechnology approach with the goal of drastically increasing biomass productivity. It garnered interest of other people in the lab, including two postdocs. The postdocs started low-level efforts to help Jill and were important internal consultants for the project. Together, we were all excited about the potential impacts of this project and eager to see Jill's project succeed. At about the same time this informal group was coalescing, a request for proposals was issued from a funding agency. Various people in my group wrote preproposals, but it was the P gene overexpression idea that proved to be intriguing to the agency and of my three preproposals submitted they invited only Jill's project for the full proposal. Jill wrote perhaps half of the preproposal and very little of the subsequent full proposal – after all, she was just a beginning masters student and didn't have the needed experience for proposal writing. Therefore, I asked the two postdocs to take the lead as co-PIs in the proposal and work with Jill to draft the full research narrative. To make our proposal as competitive as possible we intently collected preliminary data, including the attempted cloning of the P gene.

In fact, Jill had nearly cloned P several times but she had observed mutations in the sequence. Nonetheless, the postdocs felt she was on the brink of success and in a draft of the proposal they had indicated the gene had indeed been cloned. After reviewing the complete draft of the proposal about a week before it was due, I routed the proposal to all the outside co-PIs and collaborators, as well as the two postdocs and Jill.

Coincidentally, Jill was a student in my ethics course and we had just discussed the University of Wisconsin case in class in which false data had been included on a grant proposal. The current situation was too close for comfort for her. At this juncture, what were Jill's choices? She could have instigated formal charges against me and/or the postdocs for stating that P had been cloned when indeed it had not. She could have ignored the apparent falsification, which would be unethical. Instead, she did the wisest and best thing of all. She emailed me and copied the postdocs indicating her discomfort in stating that gene P had been cloned on the proposal when, in fact, the correct sequence had not yet been obtained and cloned. This was probably a difficult email to send in that the gene probably should have been cloned a few weeks earlier but there were reoccurring problems that were hoped to have already been rectified. As it turned out, the postdocs were truly hoping for the best, expecting imminent success and not intending to deceive. And I was slightly out of touch. This incident provided a point of opportunity to discuss professional ethics of proposal writing and helped me to get in touch with the true progress in the project. It also helped the postdocs understand limits of wishful thinking as it pertained to proposals. We changed the line in the proposal from "had cloned gene P" to "in the process of cloning gene P." The latter verbiage was not quite as powerful as the former, but more honest. The story ends well. We're all on the same team (still) and friends, Jill has cloned P and the grant was funded.

Judge yourself

✓ If you were in Jill's shoes, how comfortable would you be in confronting your mentor and postdoc colleagues?

✓ Are you amenable for others to confront or correct you?

Cultivating a culture of openness, integrity, and accountability

This chapter is written from the perspective of how integrity applies within lab groups, but certainly there is nothing wrong with extending some of these principles further out, such as to departments, disciplines, etc. Each lab has its own culture, and I'm convinced that the faculty member is a primary influencer of culture. I've seen labs in which the lab door is kept shut and upon knocking on the door someone inside cracks the door only slightly in response ("what do you want?"): a culture of secrecy. Secretive professors who are paranoid that someone wants to steal their research beget secretive lab members. A party professor has a party lab. An intense professor has an intense lab. No matter what the style, it is advantageous to the group to cultivate openness, integrity and accountability. How can this be accomplished? First, there must be the cultivation of trust among lab members and the boss so that everyone is working towards the same goal, namely to produce the best science possible. The graduate students and postdocs have additional goals to graduate and get great permanent jobs, respectively. The professor wants to have tenure and to be promoted. Everyone has personal goals for improvement. However, most people value being informed and enjoy open and frank communications. Most people also value fairness. Therefore, discussing ethical issues in the form of questions is powerful. Answering questions in an open fashion is empowering to everyone involved.

Openness in labs can also be encouraged in other ways. For example, why not put drafts and completed papers and proposals on a common computer where they can be accessed by all lab members, who then could provide comments and ask questions. Lab meetings in which everyone is valued and free to provide constructive criticism can also foster a culture of openness. Scientists must be willing to be held accountable for results, lack of results, and be expected to practice professional conduct. This all seems to start at the laboratory level and with the PI. I think that a culture of openness and accountability encourages informal complaints and dialogues to ensue within a group if a labmate begins to go down the wrong road. Maybe intervention happens without the boss even knowing about it. Indeed, a study of within-lab intervention recently published showed that this practice can be effective (Koocher and Keith-Spiegel 2010). In cases when a labmate performed informal intervention by confronting another labmate, the problem was corrected over 25% of the time. This rate might not seem very high, but consider the correction rate if no intervention had been performed. This is especially important considering that the intervener felt no negative fallout in over 40% of the cases and was treated with disrespect in only about 10% of the cases (Koocher and Keith-Spiegel 2010). Thus, informal intervention is constructive, especially as it leads to better accountability. Such accountability will, in turn, lead to higher integrity and better science, and we all hope, a decreased need for formal whistleblowing where there are no winners.

Judge yourself *redux*

✓ How do you handle pressure and competition? Can you stay cool or does it keep you awake at night?
✓ Do you enjoy "selling" and being the person in charge?
✓ Do you have integrity to play by the rules?

I happen to enjoy fair competition and after all these years I sleep quite well. It wasn't always so and I worried – too much. And too much worrying messes with your mind. When I played music I enjoyed being the "frontman" but I also liked just playing my instrument and letting someone else be the star. It can go either way for me. I've learned rules are there for a reason, so yes, I abide by them.

Judge yourself *redux*

✓ How would you feel to be in a situation similar to Ly and Padilla?
✓ Are you, by nature, confrontational or do you avoid confrontation?

This is so tough to be in such a position. In science, your labmates are almost like family, and so it was like, I'm sure, calling the police on your mother. I totally avoid confrontation if possible. If not, I confront.

Judge yourself *redux*

✓ Did the Goodwin debacle end the way you think it should?
✓ What are the alternative endings to the story?
✓ What would you have done differently if you were in the students' position?

No, but the current system forces this ending. I would have liked to have seen more of the students continue their studies at UW – for more support. Maybe that would happen today. A situation like this high-profile case forces change in systems. I probably would have done what they did. They tried informal intervention, which failed. Integrity demands action as they took.

Judge yourself *redux*

✓ Do you have a support network? Are there people in your lab or network that you trust to share sensitive information?
✓ Do you know your department chair and dean?
✓ Do you have the kind of relationship with your major professor that enables frank discussions?
✓ Would you ever envisage yourself as a whistleblower? How bad would an event have to be to report it?

A support network is crucial. On our own, we are bound to fail or be compromised in ways we might otherwise avoid. As a student, I did not know the chair and dean very well. There are good reasons to become acquainted with them as we see here. I would hate to be an internal whistleblower, but if that was the right thing to do, I would.

Judge yourself *redux*

✓ If you were in Jill's shoes, how comfortable would you be in confronting your mentor and postdoc colleagues?
✓ Are you amenable for others to confront or correct you?

I'm not sure I would have been as courageous as Jill when I was a student. I hope to engender enough confidence among my students to feel free to confront me. That is a lot (a lot!!) better than them reporting me to my department chair. I hope to be humble enough to take correction too.

Summary

Whistleblowing sometimes must be done to ensure scientific integrity and to correct large problems in research. It is best to systematically follow a number of steps before instigating a formal complaint; the costs of reporting should be counted. Many research issues can be handled early on and informally, which creates a climate of integrity and also strengthens a lab group to trust each other more.

Chapter 8

Authorship: Who's an Author on a Scientific Paper and Why

ABOUT THIS CHAPTER

- Scientific publishing is the main vehicle for publicly conveying scientific results.
- Authorship can be a complicated and divisive issue.
- Publishing groups have devised guidelines for assigning authorship.
- To avoid disagreements when submitting papers for publication it is best to discuss authorship in the early stages of a study if possible.

The paradigm of publish or perish is alive and well, and has matured to also include journal impact factors, citations and productivity metrics such as the H-index as a means to assess authorship and scientific success. And, for good reason, peer-reviewed papers in scholarly journals remain foundational to science itself. Until a result is actually published it doesn't really exist in the scientific canon. Multi-authored works, sometimes including dozens of authors, are growing in popularity, owing, partially at least, to expansive collaborations in science (Bordons and Gomez 2000) and the growth of multidisciplinary projects (Wuchty et al. 2007). What are the general ethics of authorship? Whose names should be included in the author list and in what order? Who gets to claim the coveted first authorship? What is the best practice in this regard? Many journals have constructed guidelines to answer these questions, but there is wide variation of adoption and philosophies among authors. While some people have attempted to make this issue totally objective and formulaic, many senior authors have their own subjective systems. There are a variety of systems (Wilson 2002) and the underlying philosophy will be discussed here as well as tips for mediating authorship disputes. The most important component of any authorship system is thorough communication with all potential authors and collaborators early in the research process.

Research Ethics for Scientists: A Companion for Students, First Edition. C. Neal Stewart Jr.

The importance of the scientific publication

Until a scientific result is peer reviewed and published in a journal, it is generally not held sacrosanct by other scientists or the public. As Macrina (2005) says so well, "Science benefits society only insofar as its findings are made public and applied" (p. 83). In fact, a good way to find trouble as a scientist is to bypass peer-reviewed publications and release scientific findings via television, press releases and other venues that are less than rigorous. Publication in peer-reviewed journals is the long-held conservative system that, while imperfect, is the *de facto* accepted method for releasing scientific results. It is only because of this fact that people who value research productivity have installed a publish-or-perish culture of rewards. Again, Macrina (2005) reiterates that scientists have a moral obligation to publish.

While scientists working in companies often don't publish (although this is changing), it is generally accepted that faculty members in most colleges and universities are expected to publish their findings regularly; being tenured and promoted are dependent on it. Winning grants, receiving pay raises, scientific prizes, and election to esteemed scientific societies are typically commensurate with number and quality of scientific publications. Bibliometric indicators of research productivity are increasingly being noticed by administrators and other people who want to measure research impacts (Bordons and Gomez 2000; Van Noorden 2010). During my tenure-track years in the 1990s, I don't recall anyone ever mentioning citations of papers as being important to research success. Now, not only do we pay more attention to how many times a paper is cited in other papers, but a cottage industry of research metrics has sprung up. Most indices and metrics are derivatives of numbers of papers, numbers of citations per paper and years publishing. An updated list of the most popular metrics along with software for computing them can be found at Harzing's Publish or Perish (http://www.harzing.com/pop.htm). Perhaps the most popular metric is Hirsch's H-index, which is defined as the number of H papers cited H times (Hirsch 2005). Let's say I published 100 papers during my career and 20 of those papers were cited 20 times or more, my H-index would be 20. Another metric, which also factors the number of years publishing is "m." M is defined as H divided by the number of years since my first article was published. So, let's say I've been publishing for 15 years and have an H-index of 20. My m-index would be 1.3. Of course, the higher the number the better, with the highest reasonable H and m being dependent on the area of science, among other factors. In my field of biology, $H > 50$; $m > 2$ are very rare, even for very senior scientists who have experienced success. $H = 20$; $m = 1$ is quite good for a mid-career and even some senior scientists. For individual researchers, H increases during a career (it is impossible for it to decrease), while m, being integrative, can go up or down. There are new bibliographic indices being created every year. The point is that numbers of publications citations is becoming increasingly important in many respects, and it is vital for new scientists to understand the games and rules of measuring research productivity (even though there is no consensus that the rules are fair or even that the correct measurements are plied).

For many of these bibliographic metrics, author order is not a factor, but I think that it might become more important in the future. It only makes sense as the first author of a paper is generally regarded as the person who has done most of the work on the project and who has primary responsibility for writing the paper. In biology, the last author is typically regarded as the person whose lab the research was performed in and the primary person responsible for the research programme. Often there are other authors besides these, and herein lies much controversy; the reason this chapter is included. In the "old days" there were many fewer authors on papers and hence, fewer authorship controversies. The more we know and need to know in science, the more people it takes to do the research to take us to the next level of understanding (Wuchty et al. 2007).

Judge yourself

✓ How do you feel about credit and fairness of credit assignment?
✓ How do you feel about being judged using metrics?
✓ Do you like to write or just do experiments and play a support role?

Who should be listed as an author on a scientific paper?

During our research ethics course we ask students to read the instructions for authors in some of their favourite journals so they can report on journal guidelines for authorship. Journal officials and scientists agree that it is imperative that the most appropriate people are given authorship and appear in the right order, but there is apparently no universal formula agreed upon by everyone to distribute fair credit. Except for the rare journal that does not address authorship at all, journals fall into two camps. The first camp uses the words "substantial contribution" and a few qualifiers to delineate who should be included in the authorship list and who should be relegated to the acknowledgments section. This first camp is pretty ambiguous as one or more of the authors determine who else should be an author by virtue of how weighty their role is deemed to be by the lead authors. The "substantial contributions" criterion can be pretty ambiguous, and purposefully so. Many editors and scientists don't want to be shoehorned into a formula. The second camp invokes guidelines developed by the International Committee of Medical Journal Editors (ICMJE). See http://www.icmje.org/ethical_1author.html. This committee has established three requirements for authorship. The first requirement can be fulfilled in one of three ways. First, an author must have made substantial contributions to (a) the conception and design of the research, or (b) collection of data, or (c) data analysis or interpretation of data. If someone is intricately involved in a study, meeting one of these first sub-requirements is typically not difficult, but again, relies on the interpretations of what "substantial contributions" really means. But at least ICMJE has sought to clarify the definition of roles of "substantial contributions." The second ICMJE requirement has to do with writing the paper or providing intellectual contributions on revisions. This step is often short circuited, thereby seemingly disqualifying many potential authors from authorship if ICMJE guidelines are to be taken literally. Here is the reason. I can't recall how many manuscripts I've sent out to co-authors who subsequently never make any revisions to the draft of the paper. "Looks great," they say. And then there is the case where the primary author(s) never sends the drafts to their co-authors. If authors don't see the paper, this makes the third ICMJE requirement also impossible to meet – that each author must give final approval to the completed paper. How many researchers, especially those who publish in ICMJE journals, take these criteria seriously when assigning credit in authorship? Glenn McGee (2007) suggests that the answer is few. He cites a study in which only three of ten non-corresponding authors met the ICMJE guidelines for authorship. He also cites problems of ghost authorship. In scientific writing, ghost authorship companies or other authors ask a prominent scientist to agree to being an author on a paper in which he or she had no involvement as a means to boost a paper's credibility or visibility. See the case study at the end of this chapter for another kind of example of ghost authorship. Providing funding or simply being the boss is also not grounds for authorship (Macrina 2005). In addition, providing equipment or lab space also does not qualify scientists for authorship (Macrina 2005). I know of senior scientists who pad their resumes by this means and it is unethical. Indeed, author inflation seems to be rampant, with many, many authors appearing on papers these days that is likely unwarranted. McGee believes that education is an important tool to address the practice of author inflation.

There is another issue that is at least, if not more damaging when assessing contributorship: assuring that people who meet authorship criteria are not excluded on the basis of status or other reasons. This, too, is unethical. "Roger Croll (1984) in his landmark paper on the topic emphasises that hourly wages, academic credit, salary, and commission are irrelevant in assessing credit. This is counter to the rationale sometimes advanced as to why technicians, consultants, and others should not be listed as authors, that is, that they have been compensated for their contribution. This argument ignores the fact that generally all who contribute to the project are paid to do so, directly or indirectly" (Pool 1997, p. 129). Indeed, there are certain organisations, such as the United States Department of Agriculture-Agricultural Research Service (USDA-ARS) that does, as a rule, exclude technicians as authors. To include a technician as an author, there must be approval by the ARS Research Leader, Centre Director or Location Director (http://www.afm.ars.usda.gov/ppweb/PDF/152–02ars.pdf). In practice, this approval to include technicians rarely occurs for some units; for others it is better. A perusal of papers coming from ARS labs clearly demonstrates that technicians rarely appear as authors, even though there are no ethical reasons given in their documents to support their exclusion. Presumably, ARS technicians perform much of the research and could easily qualify for authorship under ICMJE guidelines. Indeed, if the Research Leader, Centre Director or

Location Director agree, technicians can be included as authors, but it is certainly not automatic.

Why is it important to correctly list only the authors who substantially participate in a research project? One reason applies to most of the things we do in science: to get it right by assigning credit to those whom credit is due. As seen above, there are two kinds of common authorship mistakes. The first is not listing authors who should be on a paper. Leaving people off the authorship list when they should be listed can cause hard feelings and a sense of injustice and unfairness, even if it is an organisational policy such as we see in the USDA-ARS. Everyone knows that unjust exclusionary practices are not good for morale or productivity. It indicates that the remaining authors are getting something for nothing. The second mistake, as you might have guessed, is including people as authors who are not qualified to be authors on a particular paper. Some people might think that listing too many authors dilutes the "real" authorship. I don't know if most scientists perceive the situation that way, but I doubt it. What are one or two more authors on an otherwise nine author paper? A larger problem could be one of coercion and bullying. Just because a faculty member is on a graduate committee, or gives some materials to the student, or lets a student use a piece of equipment; these reasons do not give the faculty member the right to coerce authorship from a junior faculty member or student. Judging from stories I hear and my own observations, I believe this bullying is a serious and prevalent issue in practice. It is serious because it gives the appearance that a professor contributes more than he or she does in reality, it also fosters ill-morale, and it causes downstream injustices with regards to both students and faculty members when it comes to resource allocation and stature. I've heard an administrator express a critical concern: "I'm not sure that some of my faculty even have research programs, but I see their names on lots of papers. What does that mean?"

Because of these issues, I've become a big fan of journals, such as those in BioMed Central (BMC) that require listings of author contributions to papers. Because of the list, I can see exactly who did what in any paper. This also helps me to contact the appropriate person should I have any questions about certain details described in a paper. During tenure decisions or similar, a committee can see exactly what a person's contribution was to any specific paper.

Judge yourself

- ✓ How important is it to assure that authorship in your science is assigned correctly? How strongly do you hold your beliefs and how hard will you fight for appropriate authorship attribution?
- ✓ What are some ways you can influence best practices in authorship?
- ✓ Is it worse to include an author without good reason or exclude someone who should be an author?
- ✓ Do you prefer loose or stringent criteria for inclusion in authorship?

How to avoid author quandaries

First of all, one cannot over-emphasise or over-discuss authorship or authorship order during the course of a research project and while a paper is being drafted. In fact, delineation of authorship philosophy can begin when a grant proposal is being written, when a visiting scientist or graduate student joins the lab, during lab meetings, or even as graduate student and postdoc candidates are being interviewed. Certainly an entering prospective lab member should soon have an understanding of the publishing culture of a PI and lab. I tell everyone in my group that I want them to "own" their studies and be the first author of the research they drive. I advise them to exchange favours and work with others in the lab and outside of the lab to obtain the best collaboration and coauthorship to subsequently improve a project and its publication. My philosophy can be summed up as "the more the merrier if warranted." I tell my trainees to then put their co-authors to work in not only the experimental areas, but also in writing and revision (the ICMJE second requirement). I also advise them to put my name as the last author if they wish and put me to work too. I delegate to the first author the duties of selecting their co-authors ("and make sure you don't omit anyone") and then I'll do the final approval. It is probably not a bad idea to route the ICMJE guidelines to students and postdocs if you believe they are valid.

Of course, my style is not the only one that works and it is important to think about what your own style will be as a PI. A lab-wide system is important, and communicating the system is important to avoid problems that can eat up time and emotions. Let's imagine there are three graduate students performing related research in a lab. They all work together and understand that each one will be first author of their respective main projects. If one of them (say, Adam) decides to exclude the other two students from his paper even when the others included Adam on their papers, hard feelings will ensue if they've all met an agreed set of criteria, such as the ICMJE guidelines. These kinds of scenarios are not pretty. It is better to be inclusive (yet make your co-authors earn their keep) than to be too exclusive in authorship.

There are other authorship issues that must be negotiated. Authorship order, the journal for a submission, when to submit, and the scope of the study to be published are all items that could engender points of disagreement among authors. It might be that one person wants to publish lots of small papers and another person wishes to pool reams of data into a large comprehensive study (with that person as first author!). This decision is actually an important point when one considers citations and "fame." Only one person can be the first author on the paper (even when it is indicated that "the first two authors made equal contributions to the paper") – everyone else, including the corresponding author or PI is "et al." when there are more than two authors.

Who is listed as corresponding author is also important and is sometimes negotiated among authors. Personally, as the PI, in many cases of papers coming out of my group, I don't care if I'm corresponding author (the one who submits, deals with revisions and page proofs, and the one who pays page charges). But I'm told that some people look carefully at this as a sign of independence and being the PI. Some people consider the corresponding author to be the "brains" of a paper. So, I'll encourage my senior postdocs to be the corresponding author if they wish (though one of my grants or a program will still pay for journal charges). It is smart to be savvy and discuss these issues up front to avoid troubles when the paper is submitted. Probably most graduate students are not at the stage of their careers or experience set to be corresponding authors.

One more important consideration for getting it right with regards to authorship is in assigning responsibility and assuring accountability (Dreyfuss 2000). It is assumed that authors should be directly responsible for the overall content of a paper, whereas certain authors on multi-authored papers are responsible for specific parts of the paper. It is important to consider that your name as an author will be associated with other co-authors on both the wonderful and terrible papers. Being an author on a "paper of the year" is wonderful and a flawed and retracted paper is terrible. Guilt by association is a powerful concept. There is probably some disagreement about which authors are directly responsible when something goes wrong. I can envision cases in which not all authors can be directly responsible and accountable for mistakes, but again, we come back to guilt by association. Responsibility and accountability is worth discussing with labmates and co-authors as research and papers are developed. It is important to understand what your collaborators and labmates believe about these issues. This whole business demonstrates the importance of choosing wisely and taking seriously the entirety of each paper and the details found within.

Authorship for works other than research papers

We have delved into the rules of authorship for research papers that are to be published in peer-reviewed journals – these are the most crucial of publications. Not discussed yet are rules for conference abstracts, book chapters and review papers. In addition, it is important in some settings to discuss inventorship criteria for patents. Review papers are often invited works that review a field of study and published in a peer-reviewed journal. Book chapters are typically similar in scope and purpose and may or may not be peer-reviewed. For either reviews or book chapters, the main author invited to contribute might invite other authors to participate in writing the piece. It is assumed that all authors play important and crucial roles to the papers, but there are no rules similar to those in found in ICMJE. Conference abstracts also don't seem to have authorship rules. In many ways, abstracts and the ensuing posters or oral presentations are harbingers of future peer-reviewed publications. With that in mind, authors tend to include all the future authors of a paper, but perhaps with the order slightly altered. Thus, authorship arrangement for conference abstracts is good "practice" for preparing the authorship list and order for downstream papers. For some fields it is typical that the first author is actually giving the talk or presenting the poster, when in practice, he or she might appear somewhere else in the author order in the actual paper. The real problem often occurs in that not much thought or consideration is usually given to abstracts compared with the full paper that might follow. Sometimes a deadline sneaks up on the chief abstract author. I'm quite sure that I'm an author on abstracts I've never seen, and I've placed authors on my own abstracts in the same boat. Sorry about that! This is not the best practice. In reality, professionalism dictates that each author should see and agree with the contents of conference abstracts even though it is unlikely that all ICMJE criteria can be met at the time an abstract is submitted because of timing and logistics. For each of these types of publications, it is good to avoid authorship surprises at the end. Communication is vital.

The difference between authorship on scientific papers and inventorship on patents

Who is listed as an inventor on a patent is governed by patent law. Just because someone is an author on a scientific paper that describes an invention does not mean that the same person can legally be listed as an inventor. While authors can be determined using the ICMJE guidelines, inventors are people who are crucial to the conception of an invention. In fact a recent US court case, *University of Pittsburgh v. Hedrick* demonstrated the importance of participation in the conception of an invention (rather than showing that an invention really works) to inventorship; conception is critical. Warren and Cao (2009) conclude that the legal finding on the Hedrick case demonstrates that inventors need not know that an invention will work, but that the crucial aspect is that they have all the details of the invention worked out in their minds, which is backed up by lab notebooks or other written evidence. The Hedrick case seemingly contradicts common sense ethics discussed on publication authorship, and is therefore worth discussing.

University of Pittsburgh researcher Adam Katz and Ramon Llull conceived of a method to isolate stem cells from fat cells, which could then be differentiated into muscle cells or bone cells, or other cell types. A UCLA researcher, Marc Hedrick, joined the group at Pitt for a research fellowship for one year. When Katz and Llull filed a university disclosure, the precursor for a patent application, Hedrick was included as an inventor. Indeed, Hedrick continued his part of the research after he returned to UCLA and showed that the invention actually worked. Subsequently the patent was issued. But then, the University of Pittsburgh filed suit to remove Hedrick from the issued patent, and they won. Warren and Cao (2009) sum that, "Even though Drs. Katz' and Llull's work was scientifically inconclusive at the time of conception, the later evidence providing scientific certainty was merely a reduction to practice and not an inventive contribution." Counterintuitive with regards to publication authorship ethics, trying to include people who helped in an invention but do not fit the legal description of an inventor could lead to hard feelings and lawsuits.

Other thoughts on authorship and publications

It is especially important for young scientists to aggressively publish, be collaborative, and increase our collective knowledge in science. Most institutions value independent research programmes that yield good science as evidenced by contributions on publications. I don't object to the trend towards valuing bibliographic metrics based upon the numbers of papers authored and cited. Indeed, these help me see what research is valued and used by others. I encourage my trainees to frame their papers in such a way to garner the most citations possible, which indicates that the information is being used and spread – that it is relevant to science and society. There is a fine line between a noble approach of sharing knowledge that also acknowledges career development strategies versus doing science to get the paper published at any cost. A career is not built upon a single paper. Collegiality, sustained effort and best ethical practices are vital considerations in the long haul over a career.

Because of bibliometrics and career considerations I think it is becoming crucial to be able to identify an author unambiguously. There are efforts to assign a unique identifying number to each author. As an example, one such system is Thomson Reuters ResearcherID (my ID number is B-4709–2009). For this and all author tracking systems, the researcher voluntarily signs up. I like the idea of establishing a central system for several reasons. Other than having the ability for unambiguous tracking of authorship, such a system could be helpful in evaluating productivity of a known or unknown scientist – maybe someone wishing to collaborate with you (to answer for example, "How much does he or she publish?"), or known scientists coming up for promotion to full professor (e.g., "How much do their peers cite their works?"). There are other author ID systems being developed. Another one is ORCID developed by CrossRef, which some people think has a good chance of wide adoption.

Regardless of whether an author tracking system is widely adopted, I recommend that students and beginning scientists who have very common names take a pen name that is unique to science and use it faithfully with absolutely no variation. For example, I'm one of a few "C.N. Stewart's" around in science, but there are more X. Zhangs and D. Patel's than I can count. Popular surnames work against unique individuality in tabulating bibliometrics. But if a Xin Zhang were to add two more initials, say Y and Z, then he would have a much more distinguishing name in science, especially in any particular field. There is one "XYZ Zhang" that I found in Harzing's Publish or Perish and just one "DCB Patel." If US President Harry S. Truman – who, in fact, had no middle name – can do it, so can scientists wishing to disambiguate their name in science. If this practice is adopted, the author should always include all initials in every work. Else, it seems that we are all headed to become numbers such as a Researcher ID-type system.

Judge yourself

✓ How do you feel about being responsible, as an author, for a large multidisciplinary project? Accountable?
✓ How important is it to trust your collaborators and coauthors?
✓ How important is it to be associated with "winners?"
✓ How do you feel about your name and its intrinsic value? Would you mind taking a pen name?
✓ How would you feel about "becoming a number" through the implantation of an ID system that tracks authors in science?

Case study 1: Who is an author?

This case study is reprinted courtesy of Daniel Vasgird, PhD and Ruth L. Fischbach, PhD, MPE and the Trustees of Columbia University in the City of New York, www.ccnmtl.columbia.edu/projects/rcr

Susan Jacobs, a PhD student from a small university, sets up, as part of finishing her dissertation, a six-month internship at a prestigious larger institution in order to learn a new molecular-biological technique. Ms. Jacobs contacted the laboratory leader, Dr. Marvin Frank, a world-renowned scientist, in the hope of developing new skills for her research and also to foster a relationship with Dr. Frank, who is well connected in her field of biochemistry.

When Ms. Jacobs comes to Dr. Frank's laboratory, she is greeted warmly as a member of the team. Dr. Frank, the graduate students, the postdoctoral fellows, and the technicians include Ms. Jacobs in the weekly laboratory meetings, in which everyone participates in a free exchange of ideas about the ongoing projects in the laboratory, and which lasts for hours. In the meetings, Ms. Jacobs finds some of the ideas helpful but others less so, and gives her point of view concerning the ongoing projects. In addition, she meets weekly, one on one, with Dr. Frank, who provides significant scientific advice and one or two recommendations, which advance her work and move her in a slightly different direction. She discusses the results of her research with her mentor, Dr. Melissa Seabrook, back at her home college, by weekly e-mails and occasional phone calls, interactions that also push ahead the project she started in Dr. Seabrook's lab three years ago.

Ms. Jacobs makes great progress during the six months she spends in Dr. Frank's laboratory, and she writes a paper reflecting some important findings. Ms. Jacobs puts herself down as first author, Dr. Frank as second author, and Dr. Seabrook as last author on the paper. At the end of the paper, she gives an acknowledgment to a technician who showed her several techniques and worked with her on a few experiments.

Ms. Jacobs based her listing of authors on her understanding of the guidelines put forth by the International Committee of Medical Journal Editors (ICMJE), which say that an author is someone who has made significant contributions to the conception and design, or to the acquisition of data, or to the analysis and interpretation of data; was involved in drafting the article or revising it critically for important intellectual content; and provided final approval of the version to be published. The guidelines, which are followed by approximately 500 medical journals, say that all three criteria must be met for authorship. Ms. Jacobs would like to send her manuscript to a journal that follows ICMJE guidelines as soon as possible, because of what she feels is the importance of her results.

Ms. Jacobs gives Dr. Frank and Dr. Seabrook a draft of her manuscript for review on a Friday, hoping for feedback by Monday. Dr. Seabrook sends her comments by e-mail to Ms. Jacobs. Dr. Frank sends his comments back to Ms. Jacobs and changes the authorship listing to include Ms. Jacobs, the technician, two postdocs in his lab, two graduate students in the lab, himself, and Dr. Seabrook. Dr. Frank also gives a copy of the draft to all the members of his laboratory for discussion at the next meeting.

Ms. Jacobs is shocked that Dr. Frank added the other laboratory members to the draft, explaining to him the ICMJE guidelines and maintaining that the major intellectual and physical work in preparing the paper was done by her and by Dr. Seabrook and Dr. Frank. Dr. Frank is equally surprised by Ms. Jacobs's feelings, responding that he and Ms. Jacobs benefited from the input of all the other lab members. Dr. Frank adds that a graduate student in the laboratory, Lisa Bain, is writing a short paper that is based on some very exciting preliminary findings, and that Ms. Jacobs would be included in the list of authors. Dr. Frank says that the results of Ms. Bain's research would need further elaboration in the laboratory and that a second paper using the same data and additional studies would be more comprehensive, and that Ms. Jacobs would be included on the second one, too.

Dr. Frank insists to Ms. Jacobs that the contributions of all the laboratory members were sufficient to satisfy the ICJME guidelines for both papers, adding that the idea of a scientist acting as an independent entity is an outdated concept and that those who work around a scientist contribute significantly, helping him or her to function.

Ms. Jacobs tells Dr. Frank that she does not want to be included on Ms. Bain's paper, feeling that she did not contribute adequately. Dr. Seabrook, who follows ICMJE guidelines but was intimidated by Dr. Frank's stature, advises Ms. Jacobs not to rock the boat, to use Dr. Frank's revisions and some of the changes suggested during the laboratory review and to submit the paper to the journal with the authorship he suggested.

1. Why should Ms. Jacobs and Dr. Frank have discussed the laboratory's approach to authorship issues when she started working in his laboratory?
2. Why is the order of authorship and the listing of authors important in a research paper?
3. What is the difference between an acknowledgment and a listing as an author?
4. Although many journals subscribe to the guidelines of the International Committee of Medical Journal Editors, many do not, and many researchers do not follow the practices that it recommends. What tends to happen, and how are ICMJE standards being challenged?
5. Who among the authors takes responsibility for submitting the paper to a journal and following up with the editor and peer-review revisions?
6. What are some potential problems with Dr. Frank's submitting a paper on preliminary findings and not performing sufficient corroboratory experiments?
7. What kind of problems may arise if the same data is used in multiple papers in the research literature?
8. What might happen if someone is listed as an author on a paper for which he or she did not do any work?
9. What might have been done to resolve Ms. Jacobs's ethical dilemma with Dr. Frank about the authors on the paper?

Case study 2: The case of the ghost-writing student

Courtesy of graduate student Hong S. Moon

Dr. H. Dick Dockers is an assistant professor in a tenure-track position in a department of engineering at a major research university in California. In his statistical modelling laboratory, there is one full-time graduate student, Hanley Smyth, and one part-time student, Johnston Klub. Smyth is a PhD student in the third year of his graduate study who is focusing on developing quantitative models for automotive industry. Smyth's research project is funded by one of the major car manufacturers. The part-time student, Johnston Klub, is pursuing master's degree in Dr. Dockers' laboratory while he is working at the university in the dean's office as a staff-member.

Dr. Dockers receives a phone call from the car manufacturer's external funding officer with the bad news that the company can't continue to provide research funding, since the company is experiencing financial difficulties. Dr. Dockers has been submitting several grant proposals to various funding agencies, but with no success. The car grant is his only grant. Without further funding, Dr. Dockers can't support Smyth, the full-time student. Two years after Klub joined the lab as a part-time masters student, he was promoted to be the assistant director of university's research funding office. Klub wants to graduate soon, but his day job has made him far too busy to write his thesis. Klub explains his situation to Dr. Dockers and asks him for help. Klub implies that he may be able to provide funding to Dr. Dockers laboratory if he "helps" Klub to quickly get his masters degree.

Dr. Dockers brings Smyth in his office and tells him that his funding will be terminated soon. Dr. Dockers tells Smyth he has three options. One is Smyth writing a grant proposal to get funding for his research. And Dr. Dockers says that writing a grant proposal is not easy and takes a long time. A second option is that Smyth can find another advisor and give up his research project; this option would extend the time in graduate studies by another three or four years. In addition to these problems, none of the professors in automotive related research at his university has enough funds to accept a new graduate student, since all car manufacturers are experiencing tough financial situations.

The last option is that Smyth can "help" Klub to write his master's thesis. When Smyth asks Dr. Dockers to clarify the term "help," Johnson realises that "help" means "writing the thesis" for Klub. Smyth's reward for writing the thesis for Klub would be an arrangement for research funding to come from the university's research office for the next two years allowing Smyth to graduate. Dr. Dockers leaves the final decision to Smyth. Although Dr. Dockers mentions that last option is probably the best vehicle for uninterrupted funding for Smyth to continue his graduate program. Smyth leaves Dr. Dockers office and starts thinking about his future.

1. What is the best solution for Smyth to continue his current research and graduate?
2. "Ghostwriting" is defined as writing in the name of another. There are many commercially operated "ghostwriting" websites. Is it ethically acceptable to publish ghostwritten scientific papers?

3. There are some science fields that acquiesce to "ghostwriting." Some people argue that writing is not a part of sciences, thus ghostwriting is acceptable, unlike data fabrication or stealing intellectual property. Also they say that you can be more productive if you are working in the lab instead of spending a lot of time to write a paper. What do you think about this opinion?
4. Are there any other ethical problems in this case study?

Judge yourself *redux*

✓ How do you feel about credit and fairness of credit assignment?
✓ How do you feel about being judged using metrics?
✓ Do you like to write or just do experiments and play a support role?

Fairness of credit on papers is very important. I want the authorship to reflect who did what on a paper. For metrics and judgment, I don't really mind. Metrics can be powerful tools for self-improvement, which should reflect the quality of performance bosses are judging. I like to do it all.

Judge yourself *redux*

✓ How important is it to assure that authorship in your science is assigned correctly? How strongly do you hold your beliefs and how hard will you fight for appropriate authorship attribution?
✓ What are some ways you can influence best practices in authorship?
✓ Is it worse to include an author without good reason or exclude someone who should be an author?
✓ Do you prefer loose or stringent criteria for inclusion in authorship?

While I do feel strongly about correctly appropriating authorship, I also realise that various people have different standards for assigning authorship and the order of authors. I might question what I perceive to be mistakes but I would hope to be diplomatic. If I am the corresponding author, I might be somewhat stubborn in what I believe to be best. Communication is the key for getting everyone on the same page regarding authorship. It is much worse to exclude because of hurt feelings. This person may not be very trusting of the team in the future, so it is also a personnel morale issue as well. I am flexible with regards to ICMJE (but I like it) or the more vague "substantial contribution" model.

Judge yourself *redux*

- ✓ How do you feel about being responsible, as an author, for a large multidisciplinary project? Accountable?
- ✓ How important is it to trust your collaborators and coauthors?
- ✓ How important is it to be associated with "winners?"
- ✓ How do you feel about your name and its intrinsic value? Would you mind taking a pen name?
- ✓ How would you feel about "becoming a number" through the implantation of an ID system that tracks authors in science?

The larger the paper gets, the more conscious I feel about making key mistakes. I also need to understand all the components of a paper, whether they are large or small. Papers with fewer authors are usually more straightforward. You have absolute choice over few things in life, but you can choose collaborators and they often add coauthors; e.g., people in their labs. It is crucial to be able to trust and enjoy working with collaborators. Winning is better than losing. While I like to publish papers that will be cited a lot, I'm also happy acting as a coauthor for a less-impactful paper if it will help my collaborators and be beneficial to science. I feel like I could be identified by any number of choices of names and be happy. It might actually be fun to have a pen name. Having a number associated with my ID is ok too. It could be helpful.

Summary

Authorship is one of the most important considerations in science where positions, tenure and promotions are at stake. There is still some intrinsic unfairness in the world of authorship where ethics are not considered. As best we can, we owe our disciplines and co-workers to appropriate authorship and fair acknowledgment in papers. Issues of responsibility, accountability, research metrics and keeping track of authors are all issues that are worthy of discussion and debate. What is most important is to arrive at consensus positions regarding key issues such as authorship to facilitate durable collaborations with enhanced productivity.

Chapter 9

Grant Proposals: Ethics and Success Intertwined

ABOUT THIS CHAPTER

- Science cannot be accomplished without funding from grants.
- Granting opportunities are increasingly competitive.
- Finding funding the right way and playing by the rules sets up successful research.
- Being a fair and ethical collaborator is an important part of modern science.
- Upon award of a grant, due diligence to perform the research is an important commitment in science.

Along with publishing, funding is one of the cornerstones of modern science. Simply, without outside funding, little research can be accomplished. While grant proposals in the US most often are submitted to national agencies such as the National Institutes of Health (NIH) and the National Science Foundation (NSF), there are a plethora of agencies, companies and foundations that support science worldwide. In almost all cases, savvy grantsmanship and collaboration are necessary for funding, especially for landing the bigger grants. Best practices and ethical behaviour pave the winning path in funding research. Case studies will focus on proposal writing and collaboration. The ethics of multiple proposal submissions and doing the work before funding occurs will also be discussed, as well as financial matters. We'll examine features of being a good PI and a good collaborator.

Why funding is crucial

In "ye olden" days, there were far fewer scientists than today, many of whom were wealthy in their own right and needed no outside funding. Relative to today, nineteenth century (and earlier) science was at its rudimentary stages and was pretty cheap to do (Shamoo and Resnik 2003). Check out 1960s sci-fi movie or TV shows and you'll see "futuristic" labs appearing quite sparse, and

Research Ethics for Scientists: A Companion for Students, First Edition. C. Neal Stewart Jr.

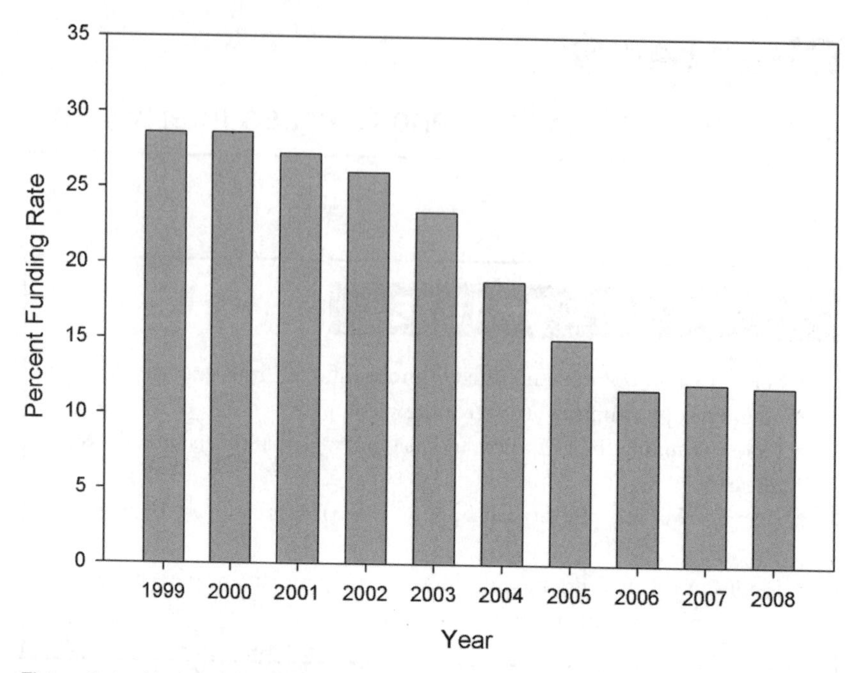

Figure 9.1 National Institutes of Health (US) grant funding rates have steadily declined from 1999 to 2006 and held steady at approximately 11–12%. These rates represent the number of awards by the number of proposal submissions (http://report.nih.gov/nihdatabook/).

Source: Reproduced by permission of National Institute of Health

mainly featuring fancy glassware containing different coloured liquid – and giant computers with tape reels. This reflected not what people thought science would be so much as what it was at the time. Other than rare "big science" projects such as the Manhattan Project to develop nuclear weapons in World War II and space-mission science in the 1960s and 1970s, most twentieth century science was relatively cheap and simple. Even 30 years ago, the entire landscape of science was vastly different than today, which included far fewer scientists and more available funding per scientist (Shamoo and Resnik 2003). Therefore, the competition among scientists for funding was less stringent and competitive than today (see Figure 9.1). As an example, in 2009, the US "stimulus package" infused new funds into science. The ARRA ploughed over $10 billion (USD) of one-time "stimulus funds" into the NIH budget that was open for competition by US biomedical and biological scientists. Every university administrator in the country actively urged faculty to go after this one-time pot of free money. As a result, a scientific feeding frenzy ensued resulting in about 20,000 applications (which also required an even greater number of peer reviewers) and, as a result, a paltry 1% funding rate. This, in turn, led to slowing down the wheels of review and award. Who knows how many wasted person hours were spent writing failed proposals that would never truly be competitive. But, what is the alternative? Funding starvation? And I kind of doubt that the 99% of researchers who were not successful during the ARRA competition simply discarded their proposals. Not at all. They'll reuse and recycle, submitting slightly revised proposals in subsequent competitions in hopes that odds will fall their way. So, the US research community will see the ARRA funding wake for at least the next three to five years, which will drive down NIH funding rates even further. It is important to note that competition for regular (R01) NIH grants has become incredibly fierce (see figure below), whereby the funding rate declined by about half between the years of 2002 and 2006, and has levelled out at around 11–12% (Figure 9.1).

Ok, so I didn't compete for any ARRA money except for a short preproposal that was turned down and also on another couple of proposals as a co-PI. My opt-out was borne out of a lack of eagerness to enter the scramble of low probability/feeding frenzy funding. But the salient facts of funding are immutable.

Fact 1: science costs money

Salary and fringe benefits for graduate students, postdocs, technical support and other people is expensive, especially in the developed world. Gone are the days where students of wealthy sponsors work for free. Materials and equipment needed for modern life science research is not cheap. And those are just the "direct costs." On top of direct costs, universities charge funding agencies money to administer grants. This is known variably as indirect costs, facilities and administration (F&A), or overhead charges. Indirect costs are used by university for utilities, clerical help, cost accounting and for maintaining facilities. In some universities and departments, a portion of these funds are shared back to departments and PIs. The indirect cost rate varies per institution and can range from 30% of direct costs to 70% or even higher. Therefore, the tough game of getting funding is made even tougher by having to budget for the indirect costs required by the home university. The typical grant size in the US ranges from $200,000 to $800,000 for many US federal agencies, which is used to fund a project for two to three years.

Fact 2: if a PI is not doing competitive science, then funding chances are low and little meaningful science can be accomplished

Bank robbers rob banks because that is where the money is. Scientists don't typically enter a field of science because they enjoy seeking funding, but to stay active in a field and produce meaningful data, funding is required. Therefore, they must write proposals and find where the money is. It follows then that given the choice of doing science in a field with little funds available or one with lots of funding possibilities and being happy either way, why not choose the easier-to-fund field? Research productivity (and enjoyability) increases when the proposal writing effort to funding ratio is low and the funding level to administration headaches is high. Spending a lot of time writing grant proposals can be tedious. Spending a lot of time writing failed grant proposals is frustrating. Writing no grant proposals and getting no money is also deemed failure and won't be tolerated by most employers at research universities. So the middle ground is maximising funding potential while decreasing efforts to get those funds. So, learning where the money is and how to get it is crucial to thriving in science. Some faculty vacillate from aiming too high, where they might continually fail, to a place where they aim too low, and acquire a low amount of funding to do research that has little real impact. The key is for each researcher to find the funding sweet spot, which, of course, will vary among researchers and fields of study.

Fact 3: the faster money is spent, the more science will be accomplished

Many scientists hate writing grant proposals, and want to maximise the time that money is held. This is a mistake. The most expensive budget line is typically for people – hiring students and scientists who will actually do the research – and so hiring and spending (within reason) is correlated with accomplishments. Of course, there are several caveats, such as hiring the right people and making sure their funding won't run out within a reasonable time (as a PI you don't want your PhD students to run out of stipend just two years into their programmes). You also don't want to create anxiety among your researchers – you want them to be focused on research. Most granting agencies do allow for "no cost extensions," giving PIs extra time to do the research. But the sooner researchers can spend the money and accomplish the goals of a project, the sooner new funds can sought for follow-up research; science should be performed in a timely fashion.

Money and ethics

Given the intensifying funding situations and the intrinsic drive for productivity that most scientists possess, many ethical dilemmas can emerge. Temptation to make ethical shortcuts is ever-present. First, getting and keeping funding is one of the most stressful things that scientists face. It was probably the *most* stressful issue that kept me awake at night when the lab was getting started and in its growth phase. Sometimes in the middle of the night, I dreamt of 20 hungry mouths to feed. It can be a nightmare, actually. But, the case can be made that without this stress there is no science. Second, to have sustained funding, not only must science be sound, but people must be treated ethically and effectively mentored. So, the theme for this chapter is that success in funding is partly the result of ethical treatment of collaborators, subordinates, administrators, and fellow scientists. That is, science funding is positively correlated with ethical practices in grants and contracts.

Judge yourself

✓ How do you generally deal with stressful decisions that could be entangled in ethics?
✓ Are you competitive?
✓ Do you enjoy planning work and then "selling" a proposal?
✓ What's more important – people or money?

Path to success in funding

This chapter is not intended as a guidebook to specific grants. Rather, the focus is in offering guidance about ethical issues that are inevitable on the route to competing for, winning and spending research funds. Integrity is key to proposal writing, grant spending and sustained success. Certainly, FFP does not spell success when it comes to grants, but then there are non-FFP nuances that are critical to maintain integrity.

Essentially, good grant getters have several common attributes. First, they all have great ideas. Second, they are able to convey those ideas and the benefits of their research to a group of peers and a wider audience. Third, they realise that success follows success (i.e., funding follows publications, leading to more of both). Fourth, people trust them with resources (money) and feel assured that PIs will do what they say they'll do. Most grant proposals are formally peer-reviewed. However, in some agencies, programme managers or officers have a lot of leeway to fund proposals and PIs they want funded. And inevitably, reviewers and programme managers tend to think of funding PIs rather than projects, even though proposals are typically project-centred. Programme managers also exist (though having different titles) in many companies and private foundations. There are even private individuals who endow or otherwise fund research at institutions. One trick of getting funded is matching great ideas with people and organisations who want to fund those great ideas. Therefore, a smart scientist will think in terms of matchmaking and goodness-of-fit to maximise research efforts and minimise frustration of writing failed proposals. Great ideas can actually carry the day, especially for young scientists with little track record of research success. So, be bold in your ideas but realistic in what can be accomplished. After that first grant is awarded and the money is spent, nearly everyone wants to know how much good science was done using the money, which leads us to back to the maxim of success following success. Funders want to know about publications and other indicators of research production, such as patents filed.

Several prior chapters, especially those on mentorship and authorship, discuss the habits of successful scientists. It is obvious that good mentorship, which leads to happy students and postdocs, which all leads to superior research, is one of the most important drivers of successful science. In addition, maintaining a good reputation as a fair-player in authorship and data acquisition and sharing shows up when proposals are peer-reviewed. No one wants to fund a louse. Best practices can make the difference between winning and losing in the grants world, and scientists should never underestimate the power of good will. I can't overstate how very thin the line is that divides winners and losers in grant competitions – a winners-take-all game. Many people, scientists included, don't like to think of research as a competition, but the fact is that a small subset in any grant competition wins the money and the others don't. Meritocracy rules, not egalitarianism when it comes to funders. Funders nearly always have a choice – they are in the position of power – especially when they are handing out lots of money, whether they want to work with a fair player who they like and respect versus a scientist with a less than stellar reputation. The choice is clear: the best science practiced by ethical scientists has the best chance for winning funding.

Fair play and collaboration

Single investigator grants are not the only choice for funding these days. A higher than ever premium is placed on scientists who play well with others to produce multidisciplinary research with high potential impact. Building teams has become more important than working as the lone hypercompetitive scientist who must control every aspect of research. Thus, it appears that there is somewhat of a dichotomy set up in the psyche of the scientist. On the one hand, the team player is rewarded. On the other hand, the competitor is rewarded. I hate to lose, whether it is table tennis or in grants. But at the same time, there is great value in team-building, collaboration, and cooperation. It produces better science and better scientists who can build upon successful teams. Many of my current grants, and the majority of my current budget, are borne of scientific collaborations. I am not the main PI of probably half the projects and half the money I'm spending. Thus, I have been sought out, trusted as part of teams, which has resulted in a relatively low grant-getting effort to productivity ratio. Other than additional opportunities for funding, there is another advantage of being a co-PI and not the PI of a grant. The PI has to expend much time and effort shepherding a proposal through all the channels, coordinating among team members, and must generally holding things together: i.e., grant administration. Being a co-PI leaves more time for science: maximising the money to effort ratio. The co-PI might not have as much control over a project, but teaming opens up avenues of science that were not accessible before and often infuses new excitement to a lab. No matter if you find yourself as the PI or co-PI, there are rules of the game that require to be followed.

Judge yourself

✓ Do you play well with others or are you more of a loner?
✓ Do you enjoy competition? A fair competition?

✓ Are you a good follower? A good leader?
✓ What is more important: getting money at all costs or heeding your moral compass?

When you are the lead PI

If you want to put together a multidisciplinary proposal and need other people to complete the team and the science, then they must be recruited (and sometimes charmed) from the inside or outside of your home institution. If you are looking for the absolute best science, often the person comes from outside your university since no university is great in all sub-disciplines. If a person is going to join your team he or she must be convinced of at least two things. First, that you will treat collaborators fairly, not only in acquiring the funding but also in doing the project, publishing the results, and fairly sharing credit. Second, that they will have a high probability of winning the funds if they enter into the collaboration with you. All your potential collaborators must be convinced it is worth their time to join your team or have you join theirs. The most important component in building trust is maintaining confidence and propriety. If you follow the Golden Rule and treat others how you wish to be treated when you are the collaborator or co-PI then you cannot go wrong.

The recipe for fair play is straightforward. First, it is important to communicate that you, as the lead PI, hold the collaborators and co-PIs in high esteem and truly need them. You must convey that without them the proposal will not likely be funded. It must be sincere. If you don't believe it, then there is no need to invite. Second, it is important to communicate that you value their input and time, and that you want to work together as a team. Again, it must be sincere. I've made the rare mistake before of treating collaborators as subordinates, which is the recipe for failure. Third, as valued team members, you will treat any privileged information and data as such and not share it without their permission. The ethics of authorship come into play in future papers together, and that concept should be communicated. It is not permissible, nor ethical, to obtain collaborators' trust, ideas, preliminary data, and then not include them as a co-PIs. It is certainly not allowable to cut collaborators out of a proposal without their permission. If someone gives data or text for inclusion in a proposal and then decides not to be a part of the proposal prior to submission, it is research misconduct to retain and use the collaborator's data or text. It is also not allowable to promise funding and then renege from that commitment. If the awarded funding is less than the proposed budget, then of course, budgets have to be trimmed. But decreased funding should not be used as an excuse to jettison a co-PI. It is not allowable to mine ideas and data from students, postdocs, and others who work for you without them deriving some benefit from the funding. Be fair. There are a lot of ethical potholes in funding. Creativity in science is important and people's ideas should be valued. Even if they leave your lab and move on to other places, they deserve authorship on papers as co-conceivers of the proposed science. They deserve proper credit. Gaining the reputation as an untrustworthy leader, tyrant, thief, or worse, is not the route to winning multidisciplinary grants and becoming a trusted collaborator. Sure, bad PIs often remain in science, but they will always land shy of their potential. They are typically not recruited for better positions because their reputations precede them. Finally, the lead PI also has to make sure people feel needed and included after the award is granted. I tell my team that getting the money is relatively easy. Doing what we promised the agency is the harder part. Since funders like to reward high performers success begets success.

It is not ok for a full professor to pressure an assistant professor for ideas that the former can take and lead as PI. The full professor should encourage junior faculty members to be independent as PIs in their own right. In general, power needs to be used with a good conscience. Administrators should not allow this kind of coercion and bullying to take place within their units.

When you are the collaborator or co-PI

So someone values your input and science – fantastic. And they are doing the hard work of managing the team and submitting the proposal. All you have to do is provide your input, do what you are asked to do for the proposal and then for the funded project. The best collaborators are responsive and do what they say they will do on time. They understand that without them the project might not have as good a chance of gaining funding. They are trusting and trustworthy to give the lead PI their best data and ideas. They don't compete against their PI with another proposal (though sometimes this happens, but full disclosure is required between teammates). A good collaborator does not wait until the last minute to send documents and needed material to the PI. A good collaborator is responsive to reporting requirements, and manuscript production after the funding is awarded. If you want to continue to be in demand as a collaborator, you must perform up to expectations and be a good team player.

Judge yourself

✓ How trusting are you? Are you trustworthy?
✓ Are you responsive and responsible?
✓ Can you work under deadlines and the pressure they create?
✓ Are you organised?

Recordkeeping and fiscal responsibility

Granting agencies, including companies, award grants and contracts so that the recipient can perform the research outlined in a proposal. In nearly all cases, the awardee is constrained to spend the money in defined cost categories and in a specified timeframe to meet the goals of the stated proposal. To spend funds on other research or in ways not specified in the proposal poses ethical and legal problems (Couzin 2006a). Spending categories can be altered such that a change better fulfils research goals and all parties know and are in agreement. The people that care the most about these kinds of things, seemingly, are auditors. Governments and universities, it seems, have no shortage of auditors. As a rule, scientists probably don't pay as close attention to money issues as they should. The money managers in institutions wish, in general, that researchers would be more careful – because of the auditors – but also because good accounting principles are tied with ethics. Since research often moves in directions that are unforeseen, it is arguable that spending funds on unforeseen items is often the most appropriate research decision. Indeed, perhaps the best decision of all is to ensure transparency about funding changes. The PI should not be perceived as being fraudulent when the reality was that scientific items took precedence over fiscal items in the grand scheme of things. That said, it is the responsibility of the PI to properly spend funds.

A young faculty member or scientist could have a steep learning curve with regards to fiscal and reporting rules. It is important to take the time to learn the rules. Ask questions of administrators and accountants. It is then very important to play by the rules. Researchers might be tempted to take the expedient or easy path and then ask forgiveness later rather than ask permission in certain cases. As the PI gets older with more at stake if ethical or legal breaches occur (e.g., a career of accomplishments), there should be a tendency to become more conscientious of accountability. No matter what the career stage, there is tremendous value in having the trust of your accountants and administrators in that their good also increases research productivity. Administrators quickly learn which PIs are eager to bend the rules and take shortcuts. Playing it straight has tangible benefits. Here is an example – one that plays out somewhere every day of the year. An administrator is contacted by a peer administrator in another university or a funder. A multi-state deal is being put together in an up-and-coming area of research. Big dollars and big research effort is on the line. The search is on for key collaborators with time deadlines approaching. Which researcher will the administrator call – the one who tries to push the rules, or the researcher who maintains a fastidious but productive research programme? Given equal or near-equal competence and fit, the administrator will call upon the most trusted researcher, and not the shortcutter who bends the rules and compromises trust.

Pushing the limits on proposals

Back to the grant proposal – the people who have the best ideas and the best grant proposal usually win. When funding rates for any particular panel typically range between 10% and 20%, the competition is fierce. Using unethical shortcuts, therefore, PIs could foreseeably be rewarded for either stealing ideas or fabricating data for preliminary data sections (or both!) when composing proposals. The line between right and wrong seems to get blurred in PI's eyes as competition (and desperation) increases. In the whistleblower chapter, we saw that situation in action in the Wisconsin case, and in the case of my own laboratory where "prophetic" data were conceived and offered in the draft proposal as already being full reality. I think many investigators try to project data into the future – to state data that will exist at the time when their grant proposals are actually reviewed. Given all the pressure brought on by competition, the practice of prophetic data might be rationalised as reasonable, but this practice is actually fabrication. In reality, I think it is actually a more effective strategy to send an update to the panel manager including any late data prior to the grant panel meeting date. This strategy shows that progress is being made, which can help a proposal. This is a more honest strategy than projecting as yet non-existent data into the future.

Stealing ideas is another way that is viewed as a strategy to get ahead. In addition to obvious ethical concerns, stealing ideas from another researcher is also a bad move for one's research career. It is inevitable that the victim of idea theft will someday review a proposal or paper (or both!) of the thief. And, recognising the thievery, the victim might not be inclined to play fair either and might see such an opportunity for retribution. This is certainly not the kind of research world we want where cascades of unethical behaviours build upon each other. A more noble victim is sure to report the bad behaviour.

It is not theft if ideas are obtained in a legitimate fashion, say from a published paper, which can be cited, or hearing a public research presentation, which may or may not be citable. In such cases, it might be more productive to collaborate rather than co-opt the idea outright to maintain good will and build a strong team. Unethical means of gaining a competitive advantage include co-opting data or ideas when acting as a reviewer of a paper or another proposal (see next chapter), or in any case when a communication is confidential and privileged. The most productive researchers I know don't worry much about theft of their own research, but many of them eagerly share stories of being the victims of idea theft. It is a good career move to play it straight with regards to proposals – cooperating and collaborating while maintaining pure competition is not easy, but it does pay off in both in the short- and long term.

Case study 1: The case of a questionable grant proposal (co-authored by Hong S. Moon)

Dr. Jackie Kohls is an associate professor in a microbiology department at an East Coast research university. He is the PI of microbial genetics lab composed of two graduate students, a full-time technician and a postdoc. He was tenured and promoted from assistant professor a year ago; a big reason was because of his several publications and obtaining an NSF grant as a PI and an Army grant as a co-PI. He does not like writing grant proposals, however. While he enjoys the process of extensive literature search and synthesising new innovative ideas, he dislikes the tedium of putting together proposals. Therefore, he asks his postdoc, who is being paid full-time from the NSF grant, to write a proposal centred around Dr. Kohls' latest idea; this proposal is to be submitted to the NIH. He tells the postdoc that Dr. Kohls will be PI and the postdoc can be the co-PI. He explains that this arrangement is the only one that would be acceptable to the university since it does not allow postdocs to submit proposals on their own. Dr. Kohls promises the postdoc that continued employment will be made more feasible if the grant is funded. In addition, he argues, the postdoc needs to learn how to write good proposals.

One day, Dr. Kohls was asked to review a manuscript that has been submitted to a microbiology journal. He reviews the manuscript thoroughly and recommends to "reject without further consideration" to the journal editor. He thinks that the manuscript is poorly written and lacks supporting data. However, he finds a nugget in the proposal that supports and expands on his own idea. He forwards the manuscript to the postdoc, along with a literature review that Dr. Kohls has recently completed. Dr. Kohls' idea focuses on the possibility of making a broad-range flu virus vaccine using a specific protein receptor. What was missing from his original idea was a specific activator – which he gleaned from the paper he reviewed.

Dr. Kohls justifies his actions of co-opting the idea by the fact that he needs another grant to keep his graduate students, lab technician, and especially his postdoc, funded and employed since his lab is due to run out of money in less than a year. He thinks that this concept and approach may eventually result in millions of dollars to him in royalties if it is commercialised in the vaccine industry. He thinks that it would not be stealing someone's idea if he uses different model organism or changes the experimental methods and approaches. Until this point the postdoc has agreed with Dr. Kohls' approach and his plan to collaborate to write the proposal. But he objects to the idea of stealing someone else's idea and begins to look for another position. Dr. Kohls is furious and tells the postdoc that there is no need to request letters of recommendation from him – that loyalty and trainee-mentor trust has been breached. After that, Dr. Kohls proceeds to write the grant proposal himself, which is indeed subsequently funded.

1. Funding is one of the most important elements in scientific research. Writing grant proposals is one of the major tasks for most PIs. Does Dr. Kohls do anything wrong with asking his postdoc to write most of a grant proposal but with Dr. Kohls as PI? Is it a misuse of NSF funds to ask the postdoc to write the follow-up grant?

2. Grant proposals are not peer-review publications. Is it ok to be less restrictive for plagiarism and intellectual property issue in writing grant proposals compared to peer-review publications?
3. You have access to someone's idea by reviewing a submitted manuscript. You review the manuscript and reject it. However, you are interested in the concept or experimental designs in the rejected manuscript. The rejected manuscript is never published in any journals, thus you can't cite a reference. What is the best way to utilise the idea or experimental designs in your grant proposal?
4. Was it wrong to give the postdoc the rejected journal submission?
5. Was it wrong to penalise the postdoc for conscientiously objecting to participate in the proposal writing and seeking another position?

Case study 2: The case of the collaboration that couldn't

Dr. Trent Wren meets Dr. Percival Pfaff at a conference. They've known about each others' work for many years, having reviewed proposals and papers. They have great mutual admiration; that is not surprising since each professor essentially has very closely aligned research interests and both are productive. Wren has the inside track for funding from a large governmental agency centre that consists of researchers throughout the country. The centre is based in Wren's institution, the University of Burd. After a few drinks after the talks, Wren explains much of UB's research proposal content and Pfaff declares that he would enjoy being a part of the centre if Wren could find something meaningful for Pfaff to do. Wren mentions that he heard a rumour that Pfaff's own university, the RAM Institute of Technology is putting together a similar centre. However, Pfaff is surprised since no one from RAM IT had contacted him about the opportunity.

After a few weeks, Pfaff ponders their enjoyable conversation over drinks. He also learned that RAM IT is not pursuing a centre proposal after all. However the same agency issues a request for proposals (RFP) for mini-centres. Pfaff figures that between himself and Wren, the two of them would be a formidable team and would wrap up all the pertinent technologies in the area. Wren thinks this is a good idea, and even though he had not heard about the mini-centre RFP, he asks if he can be the PI of the proposal. In addition, he thinks that the UB's centre (which is sure to be awarded and for which he believes he should have been named main PI) and the mini-centre, could have great synergy based upon similar approaches. Pfaff would rather be the PI of the mini-centre proposal since it was his idea to begin with. After much haggling they agree that Pfaff should be the PI. After Wren's centre director hears about this arrangement, he informs him that it might not be a good idea to join up with the RAM IT mini-centre. It could jeopardise their own future funding and set up a conflict of interest. Even though most of mini-centre proposal is already written with Wren's preliminary data being centrally integrated, three days before the proposal is to be submitted Wren pulls out of the ill-conceived partnership (flies the coop). Pfaff is livid and submits the proposal anyway – with Wren's data included.

1. What are the ramifications of including a potential co-PI's data after the collaboration is dissolved?
2. Why did Wren decide to enter, and then resign from the collaboration? Was he at fault? What drove his decisions?
3. Do you think there could be an actual conflict of interest if Wren were to be part of his own institution's (UB) centre and RAM IT's mini-centre?

Judge yourself *redux*

✓ How do you generally deal with stressful decisions that could be entangled in ethics?
✓ Are you competitive?
✓ Do you enjoy planning work and then "selling" a proposal?
✓ What's more important – people or money?

I'm getting better at detangling the pressures from what I know is right and also getting better at dealing with stress. Since I am competitive and like writing grant proposals, it certainly helps in the generation of new opportunities for funding. The more grant proposals I submit (up to a limit), the less stress I feel because experience has shown that at least some of them will be funded, thus giving my lab funding stability. I do enjoy all aspects of proposal writing. It remains a fun challenge to me. The question of importance of money or people is a crucial one. I've found that finding the right people is more difficult than finding money. If I find the kinds of scientists who fit with my group, who are bright and hard workers, then I can usually find funds to support them. In fact, they usually play an active role in funding too. People are more important than grants.

Judge yourself *redux*

✓ Do you play well with others or are you more of a loner?
✓ Do you enjoy competition? A fair competition?

✓ Are you a good follower? A good leader?
✓ What is more important: getting money at all costs or heeding your moral compass?

To confess, I'm naturally a shy loner. I've been able to overcome that somewhat in collaborations. I've always felt like I needed other people to succeed. Maybe it is an intrinsic self doubt of my own knowledge and capabilities that drives me to feel the need to collaborate. Maybe I need psychoanalysis! I like to think I can either lead or follow – I liked team sports and individual sports as a youth. And in sports, fairness is an overriding value. The moral compass and a good night's sleep wins over money at any cost.

Judge yourself *redux*

✓ How trusting are you? Are you trustworthy?
✓ Are you responsive and responsible?
✓ Can you work under deadlines and the pressure they create?
✓ Are you organised?

I think I'm trusting and trustworthy enough in science to succeed without having ensuing paranoia that someone is going to mistreat. No doubt, I've probably mistreated collaborators and have been taken advantage of, but these have been relatively rare in my career. I think that one of my strengths is being responsive to the needs of collaborators. I check email frequently and answer them. I would like to be more responsible than I am. Deadlines are a part of the reality of science. Hitting them means having the potential to succeed and missing them means certain failure. I prefer success and have become more organised to help deal with deadlines and responsibilities.

Summary

Competing for funding requires clear scientific insight coupled with ethical behaviour. Treating collaborators and competitors fairly is rewarding in its own right. Winning grants the right way is a lot of fun. Fiscal post-award ethics is at least as important as writing honest grant proposals. People appreciate having a good boss, fair collaborator and hitting deadlines.

Chapter 10

Peer Review and The Ethics of Privileged Information

ABOUT THIS CHAPTER

- Peer review is a long-established system of screening and improving research manuscripts and grant proposals.
- Peer review is based upon trust – that privileged information will not be unduly spread and used to create unfair advantage.

Peer review is amongst the most dreaded and perhaps the worst thing in science – only surpassed when there is no peer review. Indeed, some people believe that peer review is broken and needs to be fixed (Akst 2010). For now, mainly, it's all we've got for quality control. For grant proposals and publications, acceptance of a paper or project usually relies extensively on the responses of other scientists who are specialists within a particular field of study. Science is indeed a grand community in which its members are responsible for quality assurance in publishing and, to some extent, allocating grant funds. Oftentimes, the same day that I submit a paper for publication, I'm judging someone else's submission of a paper or grant. Those days scientists appreciate the smallness of the professional world that we inhabit. While altruism and the Golden Rule are often the norm, nearly every seasoned scientist has a story of how ideas and projects were "stolen" as the result of an abuse of peer review. Also, stories abound about unfair or even cruel treatment by reviewers, editors and funding agencies. Therefore, it is vital that we delve into the rules and best practices of peer review. Examining pertinent case studies will help to unveil its apparent mysteries. There will be an emphasis on the nature of privileged information. There are several current discussions and "experiments" in the worlds of granting and publishing to reform peer-review, and these will also be examined.

The history of peer review

Prior to the formation of the Royal Academy of Science in the 1660s, the medical and scientific arts existed as the work of very few people, a relative microscopic

Research Ethics for Scientists: A Companion for Students, First Edition. C. Neal Stewart Jr.

entity compared with today's gigantic worldwide science infrastructure. In those days, scientists were very loosely organised with some evidence that forms of peer review were used to evaluate the quality of research (Spier 2002). The first scientific journal and subsequent peer review system came into being with the *Philosophical Transactions of the Royal Academy of Science* in 1665. But as Spier (2002) states, until the middle of the twentieth century, there was more potential journal space than there were submitted articles, and the breadth and depth of science was miniscule relative to today. Back then, journal editors, of course, read the articles that were submitted. Frequently they would also opt to circulate submitted papers to a number of society members where submissions were informally reviewed. Incidentally, all members of each club likely were acquainted with other members. After World War II as science boomed and geographically expanded beyond old schools and cities, such an informal system was no longer workable. The numbers of scientists, journals, disciplines and their subdisciplines all increased dramatically; all were driven by increased funding to spur science competitiveness and the reach of science into society. Generally speaking, as funding increases in an area, productivity, that is, the number of papers produced, typically also increases as more scientists are trained. Sometime in the mid-twentieth century, there were more papers being generated than space to publish them, a situation that continues into today. However, it can be argued that with many journals being published exclusively online, there no longer exists any real physical space limitation for "pages," but there remains the requirement to select only qualified articles, at least for most journals. Indeed, the main function of peer review is quality assurance.

The nature of journals and the purpose of peer review

There is a virtual stratus of scientific journals. They range from "high impact" journals that publish a variety of papers that many scientists, irrespective of discipline, would be interested in reading, to geographically-centred specialty journals that fewer people read. There are numerous journals in between these two disparate strata, including very well-respected discipline-specialty journals that are international in scope. Almost all scientists want their work to be read and appreciated, therefore the higher profile journals are generally more desirable targets for publishing, but they are more difficult to pass editor- and peer-review. In practice, most scientists publish most of their research in specialty journals. Therefore, it is no surprise that the rejection rate of elite journals is quite high (10% accepted for *Science;* http://www.sciencemag.org/about/authors/faq/#pct_faq), and almost all papers are accepted in many lower-tier regional journals. Therefore, it is not surprising that the nature of peer-review varies widely among journals.

Impact factors

Journals are largely stratified according to their "impact factor." A journal's impact factor is the mean citation rate per paper per year. These are listed and compiled by

Thomson's Web of Science and grouped according to disciplines or subdisciplines. The "best" journals, those with the highest impact factors, are in highest demand for paper submissions. Therefore, editors, always wanting the best papers for their own journals, can compete for the most cutting edge manuscripts: papers in very hot fields that have profound and surprising results. Of course, some journals are too new to have impact factors and be indexed in the Web of Science or other compilations. Some journals are not indexed because they might be too obscure (regional) or published in languages other than English, the *de facto* official language of science. The alpha dog journals are aware of their stature, as are the runts of the litter. The alpha dogs want to remain at the top and the runts want to grow (Handwerker 2010).

For most of us scientists, we hope to get our papers in the best journals we can without too much haggling with editors and reviewers. Authors want papers to be sound and publishable. Some scientists want their papers to be great. No scientist wants to be embarrassed. Many scientists say they don't really care about impact factors or that impact factors don't equate to journal or publication quality. On the first charge, those who say they don't pay any attention to impact factors are either not completely honest or very naive. All serious scientists want their work read and cited (Handwerker 2010). Only dilettantes in science publish because their boss tells them to or for the purpose of simply gaining tenure.

I think just about everyone who knows about impact factors, cares about them, even if they profess otherwise. The proof is in the eating. Typically authors will submit a paper to the "best" journal (read highest impact factor) that they think will publish it. If they make the mistake of aiming too high and the paper is rejected, they will reformat and send it to a lower tier journal and so on. This is a very common practice that wastes a lot of time for authors, editors, and reviewers.

Of course, there are lots of caveats to assessing impact factors. Audience, numbers of good journals competing for good papers, and sexiness of field are among the factors to consider when judging or comparing impact factors. As an exercise, we could, for example, compare publishing in the fields of biomedicine and soil science. The top biomedical journals have higher impact factors than the top soil science journals; there are also a lot more of the former. There are also many more biomedical scientists than soil scientists to cite papers in their respective fields. Biomedical research is supported by much more funding than soil science, therefore more science per scientist is produced in the biomedical field compared with soil science. There are many more factors that could be discussed, but the important point is that within fields, there is variable supply and demand among journals. This leads to different cultures among fields and journals. And therefore editors and reviewers have various goals for reviewing papers that go beyond an assessment and assurance of quality. Many people consider the rejection rate of a journal to be an important factor of stringency and quality.

Goals of editors and reviewers

Editors of elite journals don't even send most of the papers they receive out for review. They merely reject them after a quick read since they know that most papers don't fit what they are looking for. *Science*, *Nature*, *Cell*, the *Proceedings of the National Academy of Sciences USA* (*PNAS*) are among the journals that reject a lot of papers without review. So, when a paper is sent out for review from one of these top journals, the reviewer is aware that a manuscript has passed the first step.

In fact, today I was sent a paper to review for a high impact journal. I review maybe 20 or 30 papers per year, but few papers I review were submitted to *PNAS*, *Nature*, *Science*, etc. Here were the things I looked for and thought about while I was performing the review. The main thing was to determine whether the paper is "good enough" for this elite journal. The word "good" used here means various things. Is the study going to be of interest to a lot of readers, or just those within a subdiscipline? Is the paper written so these various scientists can understand it? Is the paper a significant "breakthrough" or more of a "standard" incremental advance? Finally, and really most importantly, are the data robust enough to support the main points the authors make – high impact points, of course.

It is important to note that peer review doesn't guarantee a paper's findings are true (Wilson 2002). No matter what the journal, the reviewer's main job is to assure the data presented are sound, and of course, this assumes that the authors are being honest (Wilson 2002).

As an example of how peer review works, for the *PNAS* (impact factor = 9.38 for 2008; http://www.pnas.org/site/misc/about.shtml) paper I reviewed today, the topic was in a very exciting field, the paper claimed a very big result that would get noticed by a lot of people, but the data were not sufficiently robust to justify the claims. I recommended rejection. It is very important in big papers to make sure that there are no big holes. If this paper were to be published in such a high impact journal, the lead author's university would likely issue a press release citing that their important study was "published in the prestigious journal *PNAS*" or similar language. Science writers and non-scientific journalists would use such press releases to write popular articles about the new data and findings. People in government would take notice. Aunt Betty will read about it in her local newspaper and ask me questions about the study. Potentially, a company might license the technology, secure funding and try to make and sell products from the findings. They will use the published study as leverage and evidence that their product is genuine, useful and worth the asking price. Potentially, an environmental group will cite the study as proof that a certain technology will harm the planet. Legislators will be swayed about making new laws to protect the public. All these trickle-down effects start with the science and its ensuing publication. Therefore, "bigger" and splashier papers will have bigger real impacts and consequences. The truth is, however, most scientists publish very few high impact papers.

No matter what the apparent status of the journal, editors and reviewers should (and usually do) have very high standards for papers they decide to publish. They still sometimes make mistakes, but they rely on stringent reviewers and reviews to maintain their journals' reputations. Back to our example. Having been rejected, the authors of the rejected paper will do one of two things. Their first option is to reformat the paper and send it to a more specialised journal, maybe one that has an impact factor of 5 (still good) or less, but the published paper might not receive the recognition of the higher-impact journal. But this particular paper might still get rejected (see below). Their second option is to perform more experiments and provide a fuller picture of the result they initially claimed and resubmit to the high impact journal. This might or might not be reasonable or even possible. They might have already shown their very best data and likely the best data they can hope to collect. Sometimes things don't work out in science the way scientists hope.

This week I also received and reviewed a paper from a very good, but more specialised journal (impact factor of about 2.0). My goals and stringency for reviewing this paper were decidedly different than the paper submitted to the high impact journal. In addition, I also shared the paper with a graduate student, asking her to also perform a review on it for training purposes. It also relates to her graduate project. I asked her to keep the paper confidential. This is common practice in science, but some people think that the reviewer should get explicit permission from the editor. I never have. And to my knowledge, the training opportunities have never led to a misuse of data or breach of confidence by my trainees. That said, I have never shared a paper submitted to a high impact journal with any of my postdocs or graduate students. I suppose I believe that there might be some violation of sensibilities or the risk is greater than the potential reward. Anyhow, for the paper submitted to the specialised journal I'll ask fewer high level questions. Once published, chances are the findings appearing in the "lower journal" likely won't be repeated in the *New York Times* and many fewer scientists and other people will take notice. It will be read and perhaps cited up to 20 or 30 times in the next 10 years or so; maybe more, maybe less. It will be an incremental increase in knowledge and novel. I read the paper and suggested ways the authors might improve it. I asked for certain missing details, pointed out some typos, indicated how a badly labelled figure probably needs to be re-rendered and I also questioned a statistical method – I thought that another method would be more appropriate and useful. I gave my reasons. It was obvious that this particular submission was a significant part of a graduate student's degree research and she and the other authors will be happy when they hear from the editor that I (along with my own graduate student) and another reviewer or two recommended that the paper be published pending minor revisions. This particular journal has a 50% rejection rate. Some of these rejected papers are totally revamped and resubmitted to the same journal, and some of the rejects are reformatted, probably altered to address reviewers' comments, and then submitted to another journal farther down the impact factor food chain as discussed above.

All the above discussion about journal choice and author behaviour would seem to indicate that journal impact factor is somehow tied to the quality of science or the impact of particular papers or scientists. Most commentators about bibliographic metrics agree that this is not the case (Van Noorden 2010). Impact factor simply indicates how many times the average paper is cited per year published in a particular journal.

Which papers to review?

Editors pick and request reviews from scientists in the field. Perhaps it is human nature. I always accept the chance to review submitted papers in high impact journals, and most often review for medium impact journals and I don't typically review papers from really low impact journals (very seldom an impact factor of 1.0 or lower). I review the papers that are most interesting to me or because I want to help the editor who might be a friend. Some exceptions to my "rules" are reviewing for new journals that are in an up-and-coming area and not issued an impact factor yet, or articles that are particularly interesting to me, or for "open access" journals. Open access journals allow anyone to download published papers for free, thus I am particularly interested in seeing these journals succeed. They charge institutions and authors on the front-end to publish, but many have very good impact factors and can be viewed by anyone with access to a computer without cost. I figure that I should review about twice the number of papers per year than I publish. I rationalise that each of my papers gets sent out for at least two reviews, hence reviewing twice as many papers as I publish is minimal service to the profession. If I were to be really fair to the scientific community, the number of papers I review would be a good bit higher, since some submitted papers get rejected and must be resubmitted and re-reviewed. I spend probably an hour or two reading and reviewing a paper – sometimes longer for more complicated papers. It takes significant commitment of time and effort to perform fair reviews. No matter what the journal, their papers will never be better than the quality of the peer reviews, and so peer review is a service that scientists should take seriously.

Anonymity

Unless a reviewer wishes to state his or her name on the review, it is typically performed anonymously. Most often reviewers learn the names and affiliations of the authors before accepting the paper to review. Sometimes, I waive anonymity, especially if I think that I might be of help to the authors by identifying my-self. Some reviewers never waive anonymity, fearing retribution or some unfair treatment in the future. I've sometimes felt that way too. But, it seems to me that providing non-anonymous reviews is a way to maintain higher civility and forthrightness in the review, and a path for greater accountability. For some reason reviews can get nasty. Maybe it is the same reason why many anonymous comments to blogs, videos, etc., are now often uncivil. Accountability bears responsibility. Anyway, I don't think that a non-anonymous review system for scientific papers will catch on. Some journals experiment with various systems, such as double blind reviews (the reviewers are not told the names and affiliations of authors). There are also open review systems being explored. In such a system, everyone is non-anonymous and posts submitted manuscripts, reviews, revisions: everything on a website. The final paper will have all this other stuff with it and provide opportunity for follow-on reviews and comments post-publication, for example the journal *Biology Direct* (Akst 2010). I don't think most scientists really value that much transparency or want to wade through old reviews and revisions when reading scientific publications. I think most scientists are happy with the current peer-review system used by most journals and accept that, while flawed, it is better than any alternative we have yet conceived.

Science has typically responded to papers after they are published in predictable ways. Most papers seem to be taken at face value and used in the usual fashion to incrementally increase knowledge. Some findings are controversial. Historically, other scientists would write papers or letters to the editor to dispute certain findings they felt should not have been reported. Today, there are instances of nearly instantaneous refutation and commentary – not in journal pages or even journal websites – but in blogs and Twitter. A recent example was a paper that had stated that certain marine "extemophile" organisms could use arsenic instead of phosphorus for synthesising important compounds including DNA. The blogosphere reacted with a vengeance, which was overwhelming to the authors who did not expect the negative attention from other scientists. They especially did not expect the instant attention (Mandavilli 2011). Scientists will likely need to alter their expectations and how they respond to criticism given that blogs and the like are probably here to stay.

Judge yourself

✓ Do you enjoy critiquing papers and experiments? Are you able to offer constructive criticism without being mean?
✓ How do you feel about journal impact factors and research metrics in general?
✓ How personally do you take criticism? Can you separate your work or product from ego and self-worth?

Grant proposals

Grant proposals get peer-reviewed too. There are two big differences between granting agencies and journals. For granting agencies, they have a finite amount of money and wish to distribute it to the PIs of the best proposals. For most journals, authors simply have to meet certain criteria for publication. Yes, there is typically finite journal space, but journals don't have the hard constraints that granting agencies face with a defined pot of money to spend. Therefore, grant proposal reviewers are really looking for reasons to reject a typical proposal, say 80% or more of them. Therefore, all parts of proposals are scrutinised, including researchers' CVs, statements of facilities and equipment, and current and pending funding in addition to the proposal narrative. The second difference between paper and proposal review is that a grant panel usually wishes to reach consensus on which proposals get funded. A journal editor simply has to make an executive decision about publication and what changes need to be made to a paper.

Confidentiality and privileged information

All peer-reviewed information is considered to be privileged, confidential, and not to be exploited by the reviewer. Therefore, authors' names, ideas, data, materials and technologies divulged in a paper or grant proposal should be handled with the utmost care. Information provided in papers cannot be used or acted on until the paper is published. Information in grant proposals should always be held in confidence. Note that I do often request that my trainees assist me in providing reviews. But I request that they not further distribute materials or unethically utilise the divulged information. If there is any question about propriety, we have a discussion to clarify the ethical boundaries. As noted earlier, many scientists feel that confidence has been betrayed on occasion by suspicious timing of publication of coincidently similar work. Then again, I've observed very similar, yet equally superb papers published simultaneously in the same journal where the authors of the two groups are on opposite sides of the world. Sometimes a particular field as a whole moves in concert as great minds think along the same lines. This seems to happen with regular frequency in science where there is a complete absence of unethical behaviour on anyone's part. I'm not minimising the importance of maintaining confidence – peer review becomes ineffective when confidence is broken or when data or key concepts are leaked. Scientists get ideas from all kinds of sources – published papers, meetings, and everyday occurrences. Clearly, the line must be drawn so it is not crossed. Researchers must clearly delineate ethical guidelines before the opportunity to cheat presents itself. The Golden Rule is a good guide to keep in mind. If someone co-opted your particular idea or data that was submitted in a grant proposal or manuscript, how would you feel? If it is too close to that presented in your paper, you might feel violated or exploited. If it is the next logical step and a significant diversion or application from your own, you might not be offended at all. You might even feel flattered that something you wrote spurred on work that you would never have thought of. However, to play it safe, reviewers should never retain submitted papers as resources or any notes on the papers. Review and discard is a good rule to follow. In addition, reviewers do well to have a short memory or at least wait until their reviewed paper is published before they act on any urges to use the information provided in confidence.

Reviewers

It is almost never appropriate for a co-worker, collaborator, co-author of a recent grant proposal or paper, mentor or mentee to serve as a peer reviewer. This is a conflict of interest (see Chapter 12). Of course, there are always exceptions to some of these situations, and the potential reviewer should always make sure there is full disclosure to the person inviting the review. The point of peer review is to obtain an objective and non-biased recommendation of quality. Someone with a conflict of interest oftentimes cannot overcome an inherent bias. If I receive a paper from an editor and, say, one or more of the authors have been co-authors of mine in the past, and I still want to review the paper, and I feel like I can be objective, I'll disclose these facts to the editor and let him or her make a decision on how best to proceed. If the editor still wants me to review the paper, I'll usually submit a non-anonymous peer-review in an effort to maintain maximum transparency. Authors of papers and proposals often have the opportunity to recommend reviewers. It is poor form to recommend your friends and collaborators as reviewers. As I've served in editor capacities and realise that nepotism is attempted to be passed off, I am not very happy. I have diminished opinions about the professionalism of the communicating author, since I feel that confidence is attempted to be betrayed and objectivity subverted. I'm not a fan of being taken in a shell game.

Judge yourself

✓ Can you keep a secret? Are you trustworthy with confidential information?
✓ Do you have a short memory? Can you easily recall concepts and their origins?
✓ Do you feel desperate for ideas or do you feel as if you have more scientific ideas than you can possibly pursue?

Case study: What is responsible peer review?

(courtesy of Ruth L. Fischbach, PhD, MPE and Trustees of Columbia University in the City of New York; www.ccnmtl.columbia.edu/projects/rcr)

This case was adapted by Columbia with permission from: "Reviewer Confidentiality vs. Mentor Responsibilities: A Conflict of Interest" *Research Ethics: Cases and Commentaries*, Volume 3, Section 3 Brian Schrag, ed. Association for Practical and Professional Ethics.

Dr. John Leonard is one of very few molecular biologists working in a particular field. Dr. Leonard receives a paper to review, about a protein called survivin, which he and a graduate student in his laboratory are researching. The article was submitted by Dr. Mark Morris to *Protein Interactions*, a medium-impact journal, and the editor asked Dr. Leonard and two other experts in the field to review the paper. The article suggests a new interaction between survivin and the protein GFX and provides evidence for the fact that both proteins are necessary for the full survival-promoting function of survivin in a cell. The article also describes, though, that if there is too much survivin inside cells they die.

But the paper is fraught with problems: poor controls, inconsistent data in figures, and alternative explanations are not considered and claims are overstated. Dr. Leonard gives the paper to his graduate student Melissa Zane, who gives it a detailed critique and recommends significant revisions. Ms. Zane has never reviewed an article before, and Dr. Leonard thinks that doing so would be a good educational experience for her. Ms. Zane notes the finding about too much survivin being toxic to cells, a problem she has had working with the protein, and discusses it with Dr. Leonard. Both agree that they should lower the dosage of survivin in her experiments; the cells actually survive for a week, longer than her experience before, and then they die.

Dr. Leonard submits Ms. Zane's and his own comments about the research to the editor, suggesting that the paper be accepted only after a few more experiments are performed to validate some of the conclusions. One of the other reviewers has comments similar to Dr. Leonard's, and the editor asks Dr. Morris, the author, to make the revisions before he will accept the paper.

But in the next few weeks the interaction between GFX and survivin that is discussed in the paper remains in Dr. Leonard's mind. GFX was not a line of inquiry that Dr. Leonard and Ms. Zane were following in their research. They were focusing on other stimulatory proteins, but unsuccessfully. Dr. Leonard suggests to Ms. Zane that she add a compound to the cell culture system that stimulates the cell to produce its own GFX, a method that is somewhat different from what was in the paper by Dr. Morris that is under review. The enhancement method works. The cells live for a month.

Ms. Zane and Dr. Leonard draft a paper based on the results, which includes appropriate controls. *Science*, a prestigious journal, accepts the paper. Several months later, *Protein Interactions* publishes a revised paper from the laboratory of Dr. Morris. But after Dr. Morris sees the article in *Science* he suspects that Dr. Leonard, who was an anonymous peer reviewer on the paper, might have taken some of the ideas for the *Science* article from his paper under review. Dr. Morris knows that Dr. Leonard hadn't been working on GFX because it was hard to purify, and deduces that he used material in the unpublished manuscript to stimulate GFX activity.

1. What types of conflict of interest might arise when someone is asked to review a paper or grant application?
2. Is it ever appropriate for a peer reviewer to give a paper to a graduate student for review? If so, how should the reviewer do so?

3. Is it appropriate for a peer reviewer to use ideas from an article under review to stop unfruitful research in the reviewer's laboratory?
4. Is it ever appropriate for a reviewer to use ideas from a paper under review, even if the reviewer's method to achieve a result is different from that used in the paper under review? If so, how should the reviewer proceed?
5. What are some of the challenges in the current peer-review process, in which the peer reviewer is anonymous but the author is known to the reviewer?
6. What recourse is there for Dr. Morris if he suspects that his ideas were plagiarised?

Judge yourself *redux*

✓ Do you enjoy critiquing papers and experiments? Are you able to offer constructive criticism without being mean?
✓ How do you feel about journal impact factors and research metrics in general?
✓ How personal do you take criticism? Can you separate your work or product from ego and self-worth?

Of course, I like to help people and I also like analysing new work. I'm nicer than I used to be about it. I think a lot of reviewers fall into the bully syndrome. That is, they get bullied and then they themselves become bullies. I notice this, as a whole, more in younger reviewers. With regards to metrics, I really appreciate quantitative estimates of productivity because they help me refine research and grow. Detractors point out metrics' imperfections, which they might equate to fatal flaws. Like any metric, they can be misused, but impact factors do indicate which journals tend to have higher citation rates; helpful information. Who likes to be criticised? After being in music before science. I suffered much more rejection then as compared with now. Maybe I grew a thicker skin. The thick skin to insulate your inner person from your work is a pretty useful trait in science and life.

Judge yourself *redux*

✓ Can you keep a secret? Are you trustworthy with confidential information?
✓ Do you have a short memory? Can you easily recall concepts and their origins?
✓ Do you feel desperate for ideas or do you feel as if you have more scientific ideas than you can possibly pursue?

Secrets are easy for me to keep – at least certain ones. I don't leak confidential professional information or sensitive information. I have vices; spilling secrets is not one of them. My memory is blessedly short. I meet new people all the time. Some I've known for years. I might suffer from cryptomnesia at times though, and this can be a problem for concepts and their origins. I'm also blessed with an abundance of ideas, some say rabbits, to chase. Some are good and some are not so good, but I feel as if I have more things I'd like to work on than time or space in the lab. It is impossible to pursue all the projects I want to do.

Summary

Peer-review works best when it is played straight up with no hidden agendas – from authors, editors, or peer-reviewers. There is no place for cronyism, idea theft, malfeasance or dirty dealing in science. As authors we should value strict and stringent reviewing and meritorious decisions. As reviewers, we should value the best science, and try to influence authors and editors to publish the best papers possible. I serve as author, editor and peer-reviewer. Each role is vital to ensure sound science.

Chapter 11

Data and Data Management: The Ethics of Data

ABOUT THIS CHAPTER

- The sustained integrity of data is foundational to the integrity of science.
- Data should be archived in a way to maintain their viability and accessibility.
- Data should be made available for sharing among researchers.
- Digital data present certain challenges, especially when compiled in large amounts.
- Data wikis and other shared vehicles for data deposition are becoming more widely used.
- In many fields, sharing materials through material transfer agreements (MTAs) are as important as sharing data.
- There is a changing landscape about how unpublished data presented at scientific conferences are perceived and used.

At the centre of science is data. Except for perhaps some theoretical areas of science, all scientists think intensely about collecting, analysing, protecting, and then disseminating data from their research. Hopefully, they plan also for its preservation. The collection and analysis of data has already been covered in previous sections of this book, for example data verity and integrity. Thus, for this chapter, we will assume that data integrity is intact and thus we will focus on two big current issues in science. Our first task is to examine the landscape of archival and availability of data which would then enable their usability into the future; i.e., data preservation. Sharing summary data happens at the point of publication and the dispersal of raw data ranges from almost never in some fields to always upon publication in other fields. A vast continuum exists post-publication. Indeed, this might be *the* big issue in many areas of science now where giant data sets are easily produced. The second big issue is the simultaneous need of scientists to both present data in meetings, yet attempt to retain exclusive rights for publishing. It is interesting to look at the shifting sands of technology and ethics of walking the tightrope of presentation ethics.

Research Ethics for Scientists: A Companion for Students, First Edition. C. Neal Stewart Jr.

In the minds of most scientists, data collectively represents a sacrosanct and precious commodity – perhaps the most valuable raw resource in science. Grant proposals are written to collect and analyse it, and careers are made on publishing it. Yet, the rules surrounding the handling of data are still, at best, squishy. Despite the importance of data, there are seemingly few resources available to assure its long-term availability and accessibility. The same is true for archival research materials (Couzin-Frankel 2010). I think this is one of the main reasons that spurred the US National Academies to pen a recent volume on "*Ensuring the Integrity, Accessibility, and Stewardship of Research Data in the Digital Age*" (National Academy of Sciences 2009). The title says it all. Since data are so important (data, plural; datum, singular – but no one really ever seems to refer to "datum"), it compels us to examine best practices for us as individual scientists and the science community as a whole. We also need to look ahead to how technology will affect how data are stored, presented, and accessed in the coming decades.

Stewardship of data

I recently had a paper rejected. One of the main reasons why it was rejected was that my co-authors and I did not spell-out the data deposition plan. Indeed, one reviewer pointed out that we had actually shown very little of our original data in the paper. So, this issue hits close to home for me just now. Most funding agencies require data collected on their dime be made public. Most journals and scientists assume that when a paper is published, that data described will be made available to other researchers. How are we doing on this front?

Data of old

In the days of ancient past, read, before computers were widely used to collect, analyse and store data, there were much fewer data and their accessibility and archiving were more ad hoc than today, at least in many fields. Data were collected largely by hand, placed in notebooks or loose-leaf data sheets, and they ended up in physical files stored in file cabinets. Indeed labs still do this. All scientists (students, postdocs, research associates) students in my lab, and, I'd guess, people in most labs, are required to keep bound laboratory notebooks written in ink as official documentation of their research. The notebook pages are numbered and the notes and data are witnessed and signed off periodically by someone else in the lab. These ultimately belong to the institution. Keeping good lab notebooks is vital to scientific integrity and stewardship. Bound notebooks remain foundational for intellectual property protection – to prove that an invention happened when and what inventors may claim (see Carlson 2010). Notebooks are important also as scientific records; in fact, that is their primary importance. So, when people leave my lab, their notebooks stay behind with me and we indeed have physical file cabinets containing many lab notebooks. So, what happens to the notebooks when I move on from my current university, retire, or die? The answer is not clear. Ideally (in theory), they would remain with the university where an archivist would carefully catalogue and curate the documents. In practice, I doubt this happens much anywhere. I know of no one at my university to take care of my documents when I die. Most typical, I suspect, is that the PI either takes the most important ones to the new locale or home upon retiring. I suspect that laboratory notebooks from a career are most often eventually destroyed when the PI dies. This occurs because (as fond as we are of pontificating about the importance of data stewardship) no one wants to pay for long-term storage and cataloguing of laboratory notebooks or for the people necessary to organise the data so that they are perpetually accessible to other scientists. Scientists are still seen as mom and pop entrepreneurs with every man (and woman) for him(her)self. The same is true for scientific materials – it is difficult to archive items beyond the life of a lab (Couzin-Frenkel 2010). The hope is that most of the meaningful raw data in lab notebooks is reduced to meaningful figures and tables and published in peer reviewed literature which is readily accessible and citable. Therefore, if crucial methods are captured in publications, and if the raw numbers are averaged and presented in papers as reductionist displays, that data integrity and stewardship are sufficiently addressed in most cases, at least in fields where data are still relatively few and easily managed. Significant amounts of data are still collected and handled this way. Many journals have taken up the practice of allowing authors to include supplemental data online. This is a terrific opportunity to make data publicly available for downloading.

Judge yourself

✓ How do you feel about sharing data? Some people are data distributors ("Johnny Data Seed") and some are data hoarders. The latter feel that they may want exclusive use in perpetuity. Do you fit into either one of these two molds?

✓ How organised are you? Do you meticulously collect, record, and organise your data so that it can be shared?

Stewardship of digital data

So, there is more to my story of the recently rejected paper, which illustrates issues surrounding the storage of digital data in the field of genomics. It might help the non-genomicist to briefly review the timeline of DNA sequencing technology, the ability to find the sequence of the information – the order of the A, C, T, and G bases – that codes for life. The first DNA to be sequenced was that of a bacteriophage in the early 1970s. Until around 2005, the technology for sequencing genes had undergone a few incremental improvements. While the amount of data grew exponentially, the kinds of data output were essentially static. The original idea from the 1970s was to clone a segment of DNA and sequence it, giving read lengths per run at about 800 to 1000 bases for single contiguous molecules. Essentially, these experiments could be performed by hand

and data read visually on autoradiograms. Later on these data were collected using a more sophisticated automated capillary electrophoresis instrument, which facilitated faster data acquisition, and then analysed and stored by computer. The DNA sequence data could essentially be pasted into the laboratory notebooks. As more scientists used personal computers, and then when the internet became accessible, the US government (and then other governments) created digital data repositories; e.g., GenBank, which was started at Los Alamos National Lab. Here university, government, and industry scientists (if they wished) could deposit the sequences, functions, if known, and the donor organisms into the database. It was expected that if a researcher published a paper on a gene, that its sequence would be deposited into GenBank. All was well. But in the mid-2000s the genomics revolution accelerated as several research and development projects resulted in new instruments and technologies to generate much more sequencing data at a much faster pace (see figure 11.1). This resulted in even more DNA sequencing data being generated and added to databases. For example, in 2000, the year the first complete draft of the human genome was published, there were 8 billion base pairs of human genomic data deposited in the main databases in the US, Europe and Japan. Ten years later there were 270 billion base pairs in the system with the doubling of data every 18 months (Anonymous 2010). The technologies are so powerful now that small genomes can be sequenced in a day and the entirety of the expressed genes for more complex organisms such as plants and animals could be sequenced in a week or less as cost for sequencing has decreased 10,000-fold in the past 10 years (Anonymous 2010). Of course, it would take longer than a

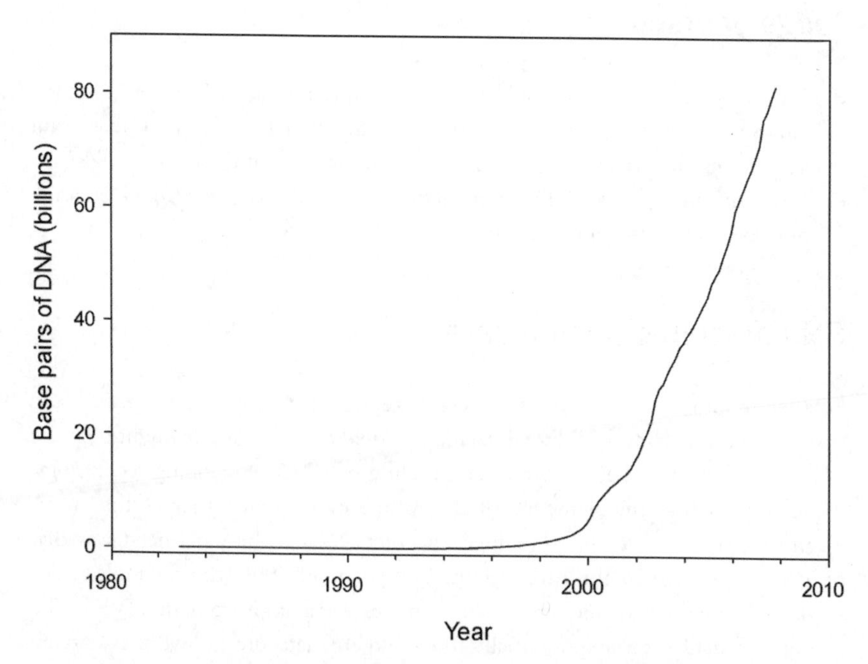

Figure 11.1 The exponential amount of DNA sequence data being deposited into GenBank (http://www.ncbi.nlm.nih.gov/genbank/genbankstats.html).

Source: Reproduced by permission of National Center of Biotechnology Information

week to analyse and make sense of the data, but the point is, whereas genes were once being sequenced one at a time, now all of them in an organism could be sequenced in parallel. The result was now millions of bases of DNA could be generated at one time. And this is where my paper had a slight hitch.

My group and now other individual (regular) labs can sequence the transcriptome (expressed genes) or the genome (all the DNA in an organism) for a plant using standard funding by sending the RNA or DNA off for sequencing and then receiving millions to billions of bases of data. So, when my rejected paper was written we assumed that we would deposit the data in GenBank or some other repository for public access, but we never stated that intention in the paper. The fact is, at the time we were actually toying with the idea of making our own database that could be expanded and organised to better serve the research community interested in the kinds of plants we study. We figured we'd make the decision after the paper was accepted; a mistake. So, for the resubmitted paper, we said we would simply deposit the data into GenBank. Biologists assume that GenBank can take all the data, *ad infinitum*, but of course data storage and retrieval is not infinite. Perhaps also complicating the review of our paper was that our collaborators and funders for the project were from a company. Maybe the reviewers suspected that we wanted to publish the summary results from the big experiment using the big data set, yet keep the raw sequence data a secret. Would that be allowable?

Data from published work and publicly-funded research must be made public

The primary duty of scientists is to collect, analyse, interpret and publish meaningful scientific results, most of which is based on data. One key doctrine of science is that science should be able to be replicated. Another is that science demands that "all the cards be placed on the table" when a paper is published. Or as the National Academy of Sciences (2009) succinctly stated, "Researchers are expected to describe their methods and tools to others in sufficient detail that the data can be checked and the results verified." It is difficult, if not impossible, to meaningfully criticise a study if data are not shown. Indeed, there are instances in papers where authors cite "data not shown" when making a point, but it is usually relatively trivial instances when it might not be appropriate or within space constraints to show particular ones in a paper. This is why supplemental information that is not on the journal's printed pages, but rather deposited on a journal's website is becoming more prevalent. Nonetheless, most funding agencies, at least in the US, require that data collected from their projects share data with other researchers and that data be placed in a public database (National Academy of Sciences 2009). But this requirement is not universal. From a stewardship perspective, I don't know of good arguments to the contrary.

In the US, the NIH has taken the issue of public access of data to another level by requiring that their grantees deposit their manuscripts that are accepted for publication into PubMed Central. Of course, many scientific publishers are not wild about this idea, because anyone can then download the paper for free and this practice potentially disrupts many publishers' economic models of business. However this public-access model is simpatico with open-access journals such as those in the Public Library of Science (PLoS) and BioMed Central (BMC) in which authors pay to publish open-access papers. Papers published in journals in PLoS, BMC, and other public-access models can be downloaded and read for free by anyone with access to a computer.

How to securely manage large amounts of digital data for accessibility

Just about everyone agrees on several important issues with regards to managing copious amounts of data. First of all, data deposition should be done. Second, the current rate of general deposition is too slow and public good is compromised as a result. Third, data sharing must take into account security of human subjects, which is a separate, yet entangled ethical issue. Fourth, new models are needed for data sharing. I described my DNA sequence deposition problem along with the problem of exponential amount of sequence data being generated and deposited into GenBank. May (2009) states that in 1989, there were 2 billion bases in GenBank, 11 billion in 2000, 45 billion in 2004 and 89 billion bases deposited by 2008. Now, other "omics" data include those from proteomics, metabolomics, ionomics, and glycomics, to name a few; then there is interactomics – how all the omics interact with one another. Large amounts of data are coming from areas outside of biology, including those from Large Hadron Collider in Switzerland and large telescopes from around the world (Lynch 2008). More data will be generated in the next decade than from last century, maybe more than in the history of the world. Clearly, new disciplines and resources are needed to manage it all.

It seems that omics and biology are leading the way in developing new paradigms for sharing. It is not an easy path. Let us assume that we can make and manage the storage capacity in the world for all the data we can possibly generate. Let us also assume that we believe it is a good idea to provide useful archives – researchers, universities, funding agencies and the public all agree. It is clear then that getting researchers within a discipline or subdiscipline to agree on standards and protocols are doable. We have seen that on several occasions – this is what scientists do best (May 2009, Lynch 2008). The key here is the development of professional standards for data deposition to assure it is of sufficient quality and an appropriate format (National Academy of Sciences 2009). This probably has to be field-specific. Bioinformaticians and computer scientists can write computer routines to help search for and manage appropriate data. We've seen the scaling of certain old routines such as BLAST to include multiple processors or even supercomputers to allow faster and more complete searches (May 2009). It is also clear that universities and academic researchers are not the ones to make and maintain large databases; that should be left up to governmental agencies.

There are instances where universities did set aside resources to maintain digital data, for example, the University of Rochester in 2003 (Nelson 2009). The cost of the repository was $200,000 USD, but if researchers refuse to use it then the resources are squandered. It is clear that large databases that are subscribed to must be supported by long term funding as well. One can make the argument that maintaining data accessibility is at least as important as collecting the data in the first place.

Judge yourself

✓ Do you procrastinate or ignore tasks that are not pressing, such as data archival?
✓ Do you think the science community should be more proactive in demanding data archival? How?

Data sharing and credit

One proposed way to handle massive amounts of diverse biological data is to create a wiki (in Hawai'ian, meaning "quick") in the spirit of Wikipedia, where researchers work together in an open environment to manage and interpret data (Waldrop 2008). Wikis are, by definition, an altruistic mechanism to foster collaboration in an open environment. There are, for example, wiki projects to annotate genes – that is to put gene function together with sequences. In one community, researchers uncover gene function information for the bacterium *Escherichia coli,* and then add it to the EcoliWiki (http//:ecoliwiki.net). At the surface, wikis seem like an effective data management vehicle when researchers are all part of a very large project and need a common meeting place for their data or for researchers with a common goal or organisms. However, as Waldrop (2008) points out, this approach might not be sustainable because of funding and focus constraints. Providing organisation for, or contributing to, a wiki does not directly lead to either funding or publications. These are the two endpoints for scientists that motivate focus. In addition, by their nature, it is also problematic to parse out credit on wikis. Therefore, wikis, while having roles in data management, might not be a cure-all for data curation.

There are various cultures among disciplines about sharing data. Some of the factors affecting sharing are historical and some are based on the nature and amount of data and collaborations. Certain fields such as astronomy and genomics have a strong culture of sharing data, whereas other fields such as social and public health sciences and climate sciences have a weak culture of sharing data (National Academy of Sciences 2009). Even if data sharing is mandated by funding agencies, there exists an honour system as such for adherence. Indeed, peer pressure to share or not to share is a strong force in specific professional communities of scientists. Some scientists wish to keep private data within large collaborations or milk as many publications out of a data set as possible; potentially setting up careers of graduate students or postdocs from their labs (National Academy of

Sciences 2009). An important aspect, therefore, in the ethics of data sharing is one of attribution of credit. Think back to the concept of egoism as described in Chapter 1. Egoism is the ethical concept that people ought to do what is in their own best interests. If there is a short-term or long-term benefit to a scientist's career in sharing data and gaining publications and recognition by sharing, then chances are the scientist will share. The opposite is true as well (National Academy of Science 2009). One way to incentivise the creation and feeding of datasets is to make them citable by issuing digital object identifiers (DOI), just as the case for all publications (Anonymous 2009; Thorisson 2009). Thus "data DOIs" elevate datasets to a standing otherwise not experienced currently – essentially making them like publications. Maybe this is the way of the future, since scientists might be able to garner credit for generating and depositing large datasets that can then be utilised by a community of scientists. But as Thorisson (2009) points out, just because datasets are made citable does not mean they would be cited. In a wiki format, where various researchers add to a database from time to time, how would credit be divvied up? Unfortunately, there are more questions than answers right now with regards to data sharing, but I would like to end this section with a personal reflection, that also takes into consideration sharing not only data, but also materials.

Sharing science, a personal perspective

My research interests include those in agricultural biotechnology and genomics. As seen above, the genomics community is accustomed to deposition of data into public databases. It is also an important task in all of our research projects. I think that biotechnology, which often generates more material than copious database-worthy data, has a different community perspective on sharing than does the genomics community. Biotechnology research generates gene constructs (e.g., plasmids) and transgenic organisms, to name two predominant materials. Other researchers periodically request these materials. Once a paper is published and the materials are described, then it is fair game to give them away – usually under a materials transfer agreement (MTA) for others to use in research. A typical academic MTA for people requesting materials to use in non-profit research is not very restrictive, containing just a few conditions. For example, the giver of the material might not want it passed from institution to institution without first notifying the source lab. The giver might want to know what the material will be used for. And the giver might wish for a certain paper to be cited when the receiving lab uses the material in a publication. The point of an MTA is to pass along interesting and useful materials, but to assure that a paper trail follows the material. For the world of for-profit entities, involving either the potential giver or receiver, the MTA business gets a little more elaborate. As an academic scientist, I am generally happy to share materials and I appreciate when others' materials are shared with my lab. And as I get older, I've become even more eager to share materials. When I was a less experienced researcher I harboured many questions about sharing materials and if I should even share in the first place.

I wondered if other researchers would do the research I would eventually want to accomplish – but do it before me. For example, will they scoop me if I send my materials? Might they use the materials for funding that I could otherwise win? Would I be enabling my own competition? Will they publish without my name on the paper or even citing my paper? Will they make my research obsolete by doing all the good science first? Will they "misuse" the materials and publish science that will make me look bad? Notice that the questions were mainly about me.

At the same time I pondered these questions, I was continually asking several other researchers for materials. Sometimes they just sent them with no MTA and no questions asked! Wow! That made life really easy. But sometimes my requests for materials were totally ignored. Follow-up emails then telephone calls went unanswered. That made my research life more difficult. In most cases, people shared their materials with me under an MTA, and that was great. It made me feel as if I belonged to a caring community. Doing the projects became easier with my group not having to reinvent the wheel every time, and I got to meet (at least over the internet) a lot of interesting scientists. At that time, I was also given myriad advice about sharing. In general my approach could be characterised as pretty cautious. I also worried about sharing my materials with two different researchers who were competitors between themselves – would that make one or both of them mad? And mad at me?

Looking back I can reduce all my worries, fears, concerns and tribulations to a simple thought. I look back with a happy heart at the many times that I shared materials and made sharing easy. I look back with regret those few times that I refused to share or gave someone the silent treatment, simply not answering their emails. Starting with the regrets: even though I had good reasons (or thought I had good reasons) not to share, I look back now with regrets that I didn't help someone I had the capability of helping. I violated the Golden Rule. I wish I could undo those instances. Of course, sometimes institutional lawyers are "at fault" and don't make sharing as easy as it should. I reluctantly accept this fact of life.

The outcome of most, but not all, the occasions I shared materials was positive. The materials were mostly helpful to people. Sometimes they even included my name on publications as they asked for my help with contributions beyond the materials themselves. I experienced new chances to collaborate with researchers I admired. Sometimes I was not included on publications on which I wish that I could have been. A couple times, the researchers gave my materials away to other people without my knowledge. Once or twice I probably enabled competitors to better compete against me. But in the end, I have no regrets about sharing – I would share all over again – I only have regrets for times of not sharing. I can reflect back on my career so far (and I'm at the projected midpoint right now), and I conclude, from an egoistic point of view, that it is good to share materials and data. Mothers probably have most the research ethics we need.

It is beneficial to science to share tricks of the trade and know-how. I arrive at the same conclusion from an altruistic point of view as well. As I think about how sharing might have affected my trainees who played key roles in developing materials I chose to distribute, I can't think of a single instance where there were negative ramifications on their careers or science. Today, I put information on how to acquire my more popular materials on my website – still in the form of an MTA with the minimum number of restraints mandated by my university administrators – but I no longer consider saying "no" for any good reason with the exception that a prior MTA disallows it.

Judge yourself

✓ How easy is sharing for you? Do you enjoy collaborations and helping people?
✓ What are the barriers, if any, to your sharing materials, data, and expertise?

The land of in-between: ethics of data presented at professional meetings

The two end points along the continuum of starting a project and publication are clear with regards to the propriety of data. At the start of a project, there's not a lot to talk about except concepts, ideas and experimental design. The end of a project or subproject culminates in a journal publication and data deposition. The middle part can get a little tricky. When you have a partial dataset, how much of it do you show and discuss at scientific meetings? What about propriety of others' data shown at meetings? The combination of enhanced competition for funds and pages in top journals coupled with advanced communication technologies lead to squishy ethical ground.

Why go to scientific conferences and workshops in the first place?

Scientific conferences can range from standing annual meetings held by scientific societies, ad hoc symposia, to special workshops to address a specific problem. The reason to attend these is to see and be seen; hear and be heard. From my perspective, conferences are wonderful places to air ideas and data, and enter into conversations with the smartest people in your field. Typically, many scientists in attendance either give talks or present posters, and they all attend talks and posters sessions. The information presented can be a combination of published and unpublished data, but typically the latter is most interesting.

Oftentimes the new data might be tentative too. Scientists can "try-out" their data to see if it passes an informal peer review of the sort that happens at conferences through normal interactions. Therefore, conferences are a combination of give and take – the Golden Rule once again. Often, talks or posters are qualified to be given at a conference by an abstract, which is a synopsis of the presentation. It might contain some data, but typically not very much can be said in a few hundred words. It serves as the advertisement and teaser for the presentation; the public face. Meetings of 50, 100, 500, or 5000 scientists are usually rewarding in different ways. Disciplined scientists attend several conferences a year in order to stay current with the cutting edge of science. After all, few scientists have the opportunity to read all the important current papers in their field of interest. Why read them, when the papers to be published next year will be squeezed into a 15 minute talk with the highlights, or a one square meter poster display.

In addition, you get the chance to meet and discuss science with the authors. It is the place to meet future students, mentors, collaborators and friends. Time at professional meetings is time well spent. Just as one example of what can happen, I recall my postdoc's poster presentation where we had elaborately laid out our grand plans for a project with some supporting data to show that the big plan was feasible or so I thought. Another scientist whom I'd never met pulled me aside and told me that his group had already tried most of what we're proposing on the poster and it didn't work. Furthermore, he said that our preliminary data were artefactual and flawed. After choking on my coffee and doughnut we had a great conversation and have been research collaborators ever since. Not only that, my postdoc could focus on better experiments instead of wasting time on things that could not possibly work. So, if meetings are such a great place to exchange ideas, what're the problems? To name a few, cameras, blogging and tweeting, which are then countered by secrecy and withholding data because of a lack of trust. Technology interfacing with ethics is behind many of the current issues that hold the power to change the face of scientific meetings.

Meeting etiquette

When I present data at a meeting, I typically want the audience to see my best and most current stuff. I want to tell an interesting story during my oral presentation, so I'll cite my published papers by including the most pertinent data to tell the first part of the story in the setup. I'll also include information generated by other scientists. For the climax, there will be my group's new data that nobody outside the lab has seen before. Taa daa (sound the flurry of trumpets)! I freely give my best story in its context to all those in attendance. Of course, there are caveats. Patentable information has to remain necessarily vague or unpresented so as not to jeopardise patent rights. If my collaborators also have sensitive information, I won't be able to share all those details there either in ethical deference to them. But generally, I don't try to hold back and I also try not to oversell (my group will chuckle at this one).

There are several reasons for this open approach. First, I want my peers to stay current with what my group is doing. They'll be among those reviewing my grant proposals and papers, so it is worthwhile to demonstrate competence and activity. I certainly don't want them to think that we're simply saying the same old thing meeting after meeting. Second, I treat the unveiling of the new data, especially those I'm not completely sure about (or their interpretation) as an audition. I want to gauge the audience reaction and listen carefully to underlying issues that might be hidden in their questions. Not too long ago at a meeting I excitedly presented results from an experiment performed by a graduate student. His interpretation of the data was questionable in my mind. I thought, sure, the reality could be exactly as the student suggested, but I questioned him about the possibility of other points of view being as likely to be valid. In the end, he assured me he was right. So, at the meeting, I played it exactly as I was coached and at the end of my talk, the hands flew up. I received very polite queries about a few things I stated, but the focus was on the potential controversy of our opinion of the data. After the session was over and as I mingled with a group of esteemed friends, where there was one person who told point-blank that he and colleagues thought I was wrong. I listened and thanked him – appreciating his forthrightness – then I proceeded to telephone my graduate student to relay the bad news given by the experts whose seasoned opinions I respected. Then, when we submitted the paper a few months later, we had written the correct interpretation thanks to an honest and helpful scientist at the meeting who were willing to give me constructive feedback. This paper sailed through review and was published. Success for both the student and science! This is exactly what I strive to learn and achieve via meetings.

Now let's imagine this scenario going a different way. Let's imagine that I'm afraid to float my trial balloon of data interpretation – either because I'm afraid of appearing inept or because someone will steal my ideas or data. Then, the chances are higher that we get hammered by the peer reviewers of the publication a few months later. I don't like that scenario, but that's ok too as long as we eventually get it right. Let's imagine that I present our data and my opinion at a conference, but it is then "stolen" via camera – my slide is photographed – and then distributed via the internet with the caption, "Stewart's wrong in stating X." Then after I find out, I'm furious and then launch into damage control mode. These are two suboptimal processes that are becoming too common of occurrences in meetings. Of course, it didn't happen that way for me, but it is not beyond possibility nowadays. And this is making some scientists more guarded about what information they choose to present at meetings.

Some ethical questions are worth asking. For example, what is the accepted etiquette for oral and poster presentations if you are an audience member? Whenever I'm in oral or poster sessions, I regularly see camera flashes. I never thought much about what happened to all those pictures. I reasoned that people photographed slides or posters for their personal use. After all, I reason, that's not much different to taking notes, and everyone would agree that nothing's wrong with taking notes during a talk. Brumfiel (2008) tells a different tale. A group of Italian researchers had collected and tightly held onto satellite data that were gauged to be important by researchers in the field of astrophysics. They had presented slides with the data at multiple conferences. They had not yet published the data, however. Apparently, the slides became predictable enough for other researchers to quickly photograph key slides at a specific point in the talk. From those slides, the picture-taking scientists in the audience recreated data and published them first, citing the Italian group's presentation in their publication. On the one hand, this seems very underhanded and sneaky. On the other hand, being slow to publish while simultaneously serially broadcasting data at meetings seems to be asking to get scooped. The Italian researchers were not very happy about the situation of getting scooped.

A recent development in meetings is blogging, tweeting and posting other's data presented at meetings on websites (Brumfiel 2009). Defenders of such actions justify blogging by saying that the whole purpose of scientific conferences is "sharing with the world what you're doing" (Brumfiel 2009). To them, it is not important if the person receiving the information is in the same room or not. I tend to disagree here. The spirit of scientific meetings is indeed sharing – but it is geographically and spatially confined sharing. The eventual published paper is intended to transcend space and time for wide dispersal – that is not really the function of a talk or poster. Conference presentations are often half-baked, after all. If blogging and tweeting my data becomes the norm at meetings, then I'll tend to present less "risky" data, and what's the fun of that? The Cold Spring Harbor conference organisers seemingly agree with my view in that they have recently instigated a policy that requires bloggers or broadcasters to get the permission of presenters prior to putting presented data on the internet (Brumfiel 2009).

I'm writing now from an airplane returning, coincidently, from a meeting in Italy. Last night, over a delicious seafood dinner and a few bottles of wine shared with old and new scientific friends, this very topic of meeting etiquette was discussed. How and when should information that is disclosed be used by attendees? One person remarked that once information is presented, it is public and anyone can use it – no questions asked. Others thought that an attempt to reference particular novel information should be made; that attribution is important. We all agreed that sharing ideas and data are enjoyable parts of meetings. There were no bloggers or tweeters at our table. But one of the scientists there told the story of someone basically publishing a book based upon lectures that had gone completely unattributed. The author led others to believe that the ideas were entirely the author's, and we all agreed that was an unethical action. We also agreed there are many grey areas, but there are some lines that shouldn't be crossed. The problem is that the lines seem to be mobile.

Judge yourself

✓ How do you feel about dispersing information presented at conferences to others not in attendance?

✓ How confident are you in your own ability to make good decisions as a presenter and attendee of presentations?
✓ How comfortable are you in sharing unpublished data and ideas?

Future of data management

We live in an era of rapidly changing norms and practices about handling data – both before it is published and then afterwards. Technology is at the root of these dynamics. First, we are able to collect more data in a single "omics" experiment now than many recently retired scientists collected in their lifetimes. The challenges to handle and archive it all are immense. Second, the internet has made accessing digital data easier than ever enabling scientists to quickly form virtual collaborations or perform meta-analyses. Therefore it is desirable to place datasets in accessible locations and in useful formats. How to actually accomplish these efficiently are not clear. Third, meeting etiquette seems to be changing as photos and blogging of preliminary results can be dispersed quickly and widely on the internet. All of these issues are certainly subjects of debate for years to come.

Judge yourself *redux*

✓ How do you feel about sharing data? Some people are data distributors ("Johnny Data Seed") and some are data hoarders. The latter feel that they may want exclusive use in perpetuity. Do you fit into either one of these two molds?
✓ How organised are you? Do you meticulously collect, record, and organise your data so that it can be shared?

I think I used to be more of a hoarder than a sharer, but that is now flip-flopped. I do need to keep my trainees in mind, however, before I give away the goodies that they might wish to pursue. They might like to have first dibs to launch their careers. With regards to data, however, scientists should make it available when they publish. My organisation has improved greatly but can still be better. I find that if I hire people more organised than me, then data can become quite orderly disposed of.

Judge yourself *redux*

✓ Do you procrastinate or ignore tasks that are not pressing, such as data archival?
✓ Do you think the science community should be more proactive in demanding data archival? How?

I don't know of any scientists who place a high priority on data archival. There always seems to be a trip, paper or proposal that takes precedence. The trick is to deal with it before it is forgotten. I wish I had the answer to the second question. Maybe we should be more proactive in creating systems that work for communities.

Judge yourself *redux*

✓ How easy is sharing for you? Do you enjoy collaborations and helping people?
✓ What are the barriers, if any, to your sharing materials, data, and expertise?

I do enjoy collaborations and sharing. The biggest current barrier to sharing materials right now is the complex nature of intellectual property and how my institution wishes to protect it. For data, there are no barriers – we just have to make sure it is published and available. Sharing expertise now in my group follows more of a distributed- than "czar" model, where the group is empowered to share their know-how as my time (and practical lab techniques) seem to be limited.

Judge yourself *redux*

✓ How do you feel about dispersing information presented at conferences to others not in attendance?
✓ How confident are you in your own ability to make good decisions as a presenter and attendee of presentations?
✓ How comfortable are you in sharing unpublished data and ideas?

I think for certain projects allowing the bloggers and tweeters to spread the presentation would be fine. For more preliminary research or sensitive data I might be more guarded. Sharing is rewarding for me even as the rules are changing.

Summary

There is a level of trust required when sharing published or unpublished data that needs to be nurtured in systems that foster openness and civility. How and where data are shared and archived are very important issues that are worthy of debate as science and technology continue to evolve. Good stewardship of data is one of the most important parts of science.

Chapter 12

Conflicts of Interest

ABOUT THIS CHAPTER

- An active and productive research programme will create a number of potential and real conflicts of interest.
- Conflicts of interest and commitment can be particularly complicated for university scientists.
- In addition to high productivity, successful scientists manage or avoid conflicts of interest.

A superb example of ethics transcending morality can be seen as we look into conflicts of interest, which can occur when someone has more than one interest; where one has the potential to corrupt the other, for which the latter is typically a constraining interest. That is to say a side pursuit corrupts the main pursuit. The underlying values, as we consider conflicts of interest, are loyalty and priority. If a person is loyal to a certain entity or cause, oftentimes the person must say "no" to other opportunities or causes that result in conflicts with the first. We see examples of needing to avoid conflicts of interest in all walks of life. In the judicial system, a lawyer cannot represent both the plaintiff and defendant, since the two sides are conflicted by nature. A gang member cannot be both a Blood and a Crip, since they compete with one another. A professor has more latitude than lawyers or gang members, but conflicts of interest can occur in academia. Lines of loyalty are valued, and it matters not whether the cause or morality in the situation is good or evil. Loyalty matters.

In science, the overriding primary loyalty is indelibly connected to discovering truth in the natural world. As we saw in Chapter 4, Herman's Second Law is "In research, what matters is what is right, not who is right" (Herman 2007). Scientists strive to uncover and truthfully explain phenomena, causes and effects without bias. It is the potential introductions of biases, both subtle and blatant, that are most worrisome to scientists. Thus, dogmatic ideology and science are not good bedfellows. However, many conflicts occur when money is involved. For example, if a conflict might arise as a university researcher performs research on a company's chemical while being paid by the company to do so. What could such a scenario mean for the subsequent publication of results?

Research Ethics for Scientists: A Companion for Students, First Edition. C. Neal Stewart Jr.

Employers of scientists must also be vigilant to guard and manage for conflicts – both for the sake of science, but also for fiduciary reasons. Of course, it is also expected that we should have solid loyalty to our employers. We must be careful not to inflict injuries to this relationship by taking on pay from others that could be construed as a conflict or compromise where none is warranted. Finally, the ultimate conflict comes when we compromise our consciences.

The dynamic landscape of conflicts of interest

Today's science differs from that of yesteryear's by the sheer quantity and breadth of possible conflicts of interest and conflicts of commitment. These are often created by commercial interests in academic scientists and their science, especially in lucrative translational and applied science (Louis et al. 1989). Although conflicts of interest abound and are probably unavoidable in certain fields, many of these are manageable. In most cases, beyond consulting, business ventures in start-up companies, patents, and the like can actually be beneficial to a scientists' home institution and its students. But on the other hand, things can go horribly wrong and out of balance when a scientist crosses a line of demarcation in the world of conflicts of interest. Case studies will be examined that involve several layers in which a scientist is forced to make decisions that have serious downstream ramifications. Potential student reactions and the effects of an out-of-bounds mentor will be discussed. We have already covered conflicts of interest regarding co-authors and collaborators and how they pertain to reviewing grant proposals and manuscripts, so this will not be discussed in this chapter. Rather, potential conflicts of a less straightforward nature will be discussed.

I will focus here on conflicts of interest and conflicts of commitment with regards to university scientists, since institutions generally have fewer rules (and greyer rules) than industry and government in this regard (and there are good reasons for it). There are many instances where conflicts can crop up. Why should we not worry about conflicts for industry and government employees here? For industry scientists, the rules are clearly understood that the conflicts' world is black and white. The company that employs a scientist expects no outside interests for that scientist, at least none where there is opportunity for payment or loyalties to be potentially compromised. Whatever is done at the company by company employees is owned by the company. In many ways, the situation is the same for government employees. A few years ago, the National Institutes of Health employees could perform outside work as consultants, but that is no longer the case, because of conflicts of interest that ensued that seriously affected the NIH to accomplish its mission. The relationships that industry and governments have with their employees is described by high fidelity – no chance to play the field. In exchange, compensation and/or employee benefits are motivators to remain loyal and conflict-free.

Potential conflicts of interest for university scientists

Let's clear up a potential misunderstanding for the lead-in discussion about the differences between industry/government and universities. I'm not advocating that it serves universities or science for faculty members to be disloyal in any way. And I don't think that professors ever want to be viewed as disloyal to their employers. I think that sometimes it is simply a slippery slope driven by prospects of earning additional money or the pursuit of compelling outside interests that cause faculty members to find themselves in conflicts of interest dilemmas. In some cases, maybe most, these pursuits can be beneficial to professor, student, university and society. Today's universities want their faculty to be interesting, in demand, and entrepreneurial, but they also need to maintain appropriate boundaries to prevent the university and objective science missions from being compromised. To execute a university's mission of educating students, providing service for stakeholders, and creating knowledge, it is sometimes advantageous to enter into agreements and situations that are somewhat muddy. For example, faculty members often act as industry consultants, expert witnesses and serve on advisory boards. Within reason, these are opportunities that universities encourage their faculty to serve. All outside relationships should be disclosed or disclosable at the appropriate time to the primary employer. I use the "newspaper rule" as my own guide. That is, if a relationship or deal involving me were to show up on the front page of my local newspaper, how would I feel? Would I feel embarrassed and wish to hide it from my employer and colleagues, or would it be the reverse? Conscience is a good guide. So is the newspaper rule.

I am currently a professor at a public university. I teach a few courses and run a research programme that has been funded by US federal agencies, companies, foundations, an endowment, and various other university funds. By far, most of my research funds are from US governmental agencies. Funders and the university officials expect me, first and foremost, to be productive in these educational and research pursuits. I share these expectations and values with them. University administrators would not be thrilled to hear news of my being slack, nor would they be pleased with a drop-off of funding or publications. They want to hear that students are benefiting from my teaching and research programmes. Staying productive and close to the cutting edge of science is also important for my career and competitiveness. Thus, our goals match. That said, I've also founded and participated in university start-up companies using university – owned intellectual property. I've been paid as a consultant and expert witness by companies. I've been paid to speak and serve on grant panels and government agency scientific advisory panels. I also serve as a member of a scientific advisory panel for a company. Are any of these conflicts of interest or commitments? Are they allowable or should aspects of these activities be mitigated? The big questions: Are they good for universities and for science and society? Is there potential for harm? Is any harm realised? These are all important questions for me to ask myself and the university to ask as well.

We'll examine what constitutes and defines conflicts of interests and commitments and try to grade them on severity. We'll also grade their risks and benefits. One handy tool that I'm using as an outline and starting point is my university's outside interests disclosure form. Most colleges and universities use these instruments to flush out yearly disclosures about faculty activities and compensation outside of the institution for each of their faculty and staff members. Collectively among institutions, university lawyers and administrators have thought a lot about what potential conflicts should be disclosed. One driver for requesting disclosures is that in every university there have been cases of faculty members who have gone overboard on certain outside activities or who have certain interests and have embarrassed or compromised the university and its mission in the past. See the newspaper rule.

Conflicts of interest vs. conflicts of commitment

Conflicts of interest deal with the potential of a lower cause to corrupt a higher, or constrained cause. We can think of an outside activity as a potential source of pollution.

They are usually defined as professional vs. private interests where someone could have some personal or financial gain outside of their primary professional duties that could impair university performance or corrupt scientific objectivity. In contrast, a conflict of commitment might have more to do with a time or personal energy allocation that could serve to diminish full focus on the primary mission of a scientist and faculty member. One can imagine several instances where there would exist conflicts of commitments, but with no conflicts of interest. For example, a deep involvement doing *pro bono* scientific research for a good cause might incur a cost to one's primary job, which is teaching and doing funded research to support students and university science. In this case, there would be no personal or financial gain, but it could result in an inappropriate time sink, which distracts a professor from his or her main duties of research and teaching. These two entities, conflicts of interests and conflicts of commitment tend to get rolled into one entity for practical purposes. Below we'll cover the main conflicts that university scientists face.

Outside positions

Most universities are not keen on their faculty having political or for-profit positions outside the institution. For the latter, some measure of these are typically tolerated, but there must be clear delineation between duties as a faculty member and duties in the private sector and how time commitments are to be managed so that faculty duties are not compromised. Typically, institutions look at three things. First, they want to know how much money is involved, say, in terms of salary per year if it is greater than some defined threshold, say $10,000 per year. Second, they want to understand how much time commitment is required for the outside interest in terms of days per month. For example, there is a big difference in becoming the CEO of a company that has one employee (you) that deals with trading baseball cards, a pursuit that can be accomplished on the weekend, and becoming the chief technology officer of a start-up biotechnology company that has 10 employees. This leads us to the third concern: the nature of the business of the outside interest. Universities don't like to be in competition with their own employees. Neither do they relish losing a valued employee to the private sector.

Consulting

Many university professors consult for companies in their area of expertise. This is typically a win-win-win for universities, faculty members and companies. Institutions become concerned when the sheer volume of consulting becomes too great or the annual income from a single consultant goes over a certain threshold. Typically, universities will allow professors to consult a few days, say, two to four per month as part of their official duties, leaving the exact details up to their individual professors to work out. They view this moderate amount of consulting as a worthwhile and manageable endeavour. For example, because of my own consulting work, new contracts and grants have been initiated for the university and students have been provided opportunities for employment and professional and scientific growth. Improving economic development and impact is part of a university's mission. Of course, if the compensation for consulting becomes too high, university administrators might question loyalties. They might also question whether there is a conflict of commitment, especially if trainees or colleagues ask the head of department if they've seen a certain missing person recently.

Boards

Being on a scientific advisory board of a company or steering committee for a non-government agency, or even a panel for a government granting agency is seen as beneficial activities for universities and faculty alike. Many of the same reasons apply as seen above for consulting. However, serving on boards are viewed as more long-term commitments, and the rules of engagement are somewhat different because of potential longevity and depth of involvement. I personally have no worries about having consulting agreements with two or more companies in the same research discipline, even during the same overall time span (but not on the same days). Companies typically request I sign a confidentiality agreement that entails my not sharing compromising information with anyone else. I bill by the hour and so my loyalties to each company, time wise, are fixed, discreet, and discrete. Being on a board is both a deeper commitment, with regards to information shared and it is typically also of a longer duration – years instead of days for consulting. It is typically also made public from the company's perspective. Therefore, the professor must be careful not to develop conflicts of interest with regards to serving on two boards for companies who are in the same field. Boards can also be tricky with regards to the amount of time needed to serve; it can potentially conflict with time needed to do research and teach. But again, boards can lead to new synergistic opportunities that can benefit universities. Often alumni, even undergraduate students, later create technical or scientific companies, whereby they tend to contact a trusted professor to consult and later serve on a board of advisors. Universities welcome the maintenance of contacts with their alumni who have a sense of loyalty to the institution. Again, it could mean funding and job opportunities for current students and increased visibility for the institution.

Start-up companies

Start-up companies are usually founded by faculty members to develop products stemming from research performed in an academic lab and perhaps patented by the university. A professor, typically with tacit approval by the university, might judge that a viable path to economic development and commercialisation would be to create a company rather than simply out-licensing the invention to an existing company. In this way, the inventor can shape its development in a stronger way. There are other reasons to found companies too, including potential sizable financial benefits to faculty members, universities, and their surrounding communities. In fact, entire industries can be founded and regions of excellence by university start-ups. The San Francisco Bay and Boston are two examples of biotechnology hot-spots as a result, at least in part, from companies co-founded by faculty members, e.g., Genentech and Biogen, respectfully. Universities can own equity in companies and stand to greatly benefit if companies succeed. But at the same time, start-ups are also potential pit-traps for conflicts. For example, a conflict of interest might arise if the professor chooses to have line management duties – for example as the CEO. It is cleaner if the professor has scientific or technical oversight duties in which there are no people-management responsibilities; that is, the scientist advises the science and not supervising company employees. It is ok for professors, or even students, to have equity stakes in start-up companies. It is fine also for professors and students to benefit from patents. Conflicts of interests come in to play when the equity stake for the faculty member (and university) is too great; even far less than controlling interest can be considered too great. For example, Stanford University will never own any more than 10% of any of its start-up companies. Perhaps the bigger potential problem surrounds conflicts of commitment. Start-up companies are a lot of work to launch and then to sustain. If the faculty member is spending so much time to be the CEO, then it is not hard to imagine teaching and research responsibilities might suffer. There is also the potential problem of student or postdoc exploitation. Students paid from university grants should not be expected to work for the start-up. There should be clear lines of delineation and it is smart to keep department heads and deans in the loop with regards to managing conflicts.

Financial interests

Universities might be concerned about outside financial interests that are related to a professor's expertise. There might be a temptation to use the university position for large financial gain or even for the exploitation of university students or employees. Let's say that I have sizable stock holdings in an agricultural company that does business in my particular area of expertise. It is not incomprehensible that I could be tempted to alter data or shade findings in my research on a blockbuster insecticide to drive up stock prices to my direct financial benefit. Most professors make it a point to stay away from these situations because of this exact sort of risk. Faculties' reputations are the foundation of their credibility in their fields.

Family's interests

Most of the above items also apply to the interests of a faculty member's immediate family: spouse, children and parents. Again, this is a precaution to stave off temptation of direct gain because of corruption.

Judge yourself

- ✓ Are you susceptible to temptation or greed? Are you especially motivated by money?
- ✓ Do you have diverse interests that could be targets for potential conflicts of interest or commitment?
- ✓ Some people have idealistic delineations that lead them to completely separate science and universities from companies and profit-motives? Do you hold these ideals?

Conflicts of interest within labs or universities

Many scientists enter into labs and universities holding pure scientific interests, ideas and ideals. Given that there is a good bit of stochasticity governing where we study, with whom, and what science we decide to pursue, there are many choices about the details of what we do in science. That said, during the early parts of our training and careers there are not many opportunities for the conflicts of interest to occur. Graduate students, postdocs, and assistant professors are innately focused on graduating, finding a permanent job and getting tenure, respectively. There is not a lot of time for outside activities. Also, these opportunities come when a scientist has a certain amount of experience and expertise.

That said, conflicts of interest internal to lab groups can involve new and experienced scientists alike. Bluntly stated, it is generally not a good idea to directly compete with your own lab members or people at your university. Or your boss! Or your ex-boss. Yet these kinds of competitions seem to happen as frequently as best friends discovering that they often dress like each other. It is almost as if randomness is not so random after all. In the pursuit of publications and success, these internal conflicts are not very rare, but it takes energy to avoid or manage them.

There is no right answer to the dilemma of finding that you are in direct competition in science with someone that would ideally be a collaborator. The two people you definitely don't want to compete with are your advisor and former advisor. This is like dating your best friend's old girlfriend. It almost always ends badly with ill-feelings that can compromise both parties' effectiveness in science.

Beyond the advisor, there are many grey areas. The optimal, although at times idealistic and not feasible, approach, is to face all potential competitors directly and lay your cards on the table. Hopefully, the competitors will do the same. I've formed collaborations with former competitors. I have also decided to not pursue research that I knew might put me in a sticky situation with someone I respect and admire. Again, there is no one right answer. The best thing to do is approach a situation with integrity and dignity and to do what's best for science, your interests and career. It often helps to think long-range in these matters, and not just about the degree or next paper.

Another way to handle impending competitions is to *not* speak about the topic of interest to people who you think might be working along the same lines of research. In this way, not knowing all the facts prevents you from being "contaminated" with ideas that could be used unethically. There have been a few instances where I've decided to guard against contamination rather than putting all my cards on the table. It was not that I was afraid of being scooped. It was that I was concerned about knowing what someone else was doing that could intractably affect the course of my own project, and I didn't want to chance it. Here is an example. I participate in a large project in which a key early activity in the grant was to decide which genes to investigate with regards to function for cell wall biosynthesis with an eye toward wall degradation. I was invited to review several grant proposals on this very topic during a key period of time when our project was in the early stages of deciding which genes to study. I declined to review all proposals having to do with cell wall genes. Why? They could have invariably shaded my judgment one way or the other. I could have been tempted to use the information to benefit my big project. But my primary concern was for the PI of each grant proposal I'd review. If I were the PI, I wouldn't want to provide crucial early information to a huge centre that could potentially scoop me. My other concern was for the centre. It seemed to me that the fairest way to play this situation was to remain uncontaminated. Once the centre had its big gene list underway I felt I could return to my normal activities of reviewing related proposals without fear of inadvertently acting inappropriately in what could have been a no-win situation.

Judge yourself

- ✓ When it comes to ideas and projects do you have almost compulsive or obsessive tendencies that could shade your view toward unhealthy competition?
- ✓ What is your style with regards to confrontation and compromise? These are important conflict management tools.

One last issue to discuss is intralab competition. When the PI is narrowly focused on a particular topic, there will invariably be aspects of the research that overlaps among the interests and pursuits of graduate students and postdocs. The goal for lab personnel and especially the PI is to minimise these overlaps. However, some wayward PIs (yes, alas, I've done this before) assign the same project or close-to-the-same project to more than one person as a bet-hedging strategy or to set up a "friendly" competition with the goal of completing a project at seemingly a faster pace than if one person had sole ownership of the project. The winner is then feted with first authorship. This play often turns out bad because there is a loser, and the loser isn't always happy. The loser's friends might not be happy either. In fact the entire lab (even the winner) could be unhappy because of the lousy morale that ensues during and after the competition is completed and the winner collects the reward. Therefore, intralab competition should be minimised for the sake of total productivity and integrity of management. The PI has the most control of this situation to assure that most, if not all, lab members are winners.

Case study 1: The case of the crowded room – who should investigate a line of research when many people have an interest in it.

PhD student Ginger Gonzales was recruited to a large university in the Midwestern US to work in the lab of Dr. Miles Pinto. Dr. Pinto is a productive PI, having multiple grants from the NSF and NIH. In addition he is a Howard Hughes Medical Institute Investigator, which gives him sizable funding per year to pursue essentially any line of research he wishes. Dr. Pinto sees great potential in Ginger and he recruited her to his lab to develop and pursue the project of her choice. Dr. Pinto's interest is broad, but he is best known for applying ecological principles to epidemiology and using various organisms to model and better understand epidemiological patterns. Ginger decides to focus on plants and ephemeral epidemiological phenomena in a relatively simple high latitude peatland system. She also wishes to incorporate global climate change parameters into her modelling and empirical experiments. Since this is somewhat out of Dr. Pinto's direct experience (although he is fascinated with the potential of the project), he encourages her to speak to several faculty members within his department of biological sciences and also in other, more applied, departments.

As Ginger talks with many people at the university about the project, she gets helpful feedback from various faculty members. They also are encouraged by her initiative and ideas. One way faculty can help students is simply by engaging in scholarly and open discussion about ideas. Another way is volunteering the use of lab equipment and materials with few-to-no strings attached. A third way is to volunteer to serve on her committee, which a couple of professors do. She contacts scientists at other universities and also gets cooperation and good ideas from some, while a few others are more standoffish. When she asks Dr. Pinto about the different responses, he tells her that most scientists, especially close colleagues, respect Pinto and his group and are eager to see science progress. They especially enjoy helping students. Pinto explains that many good scientists have more ideas than they can pursue and are excited to help generate other ideas that others can follow-up to do good science. Perhaps the more closed responses from others outside their university signify that she could have potential competitors for her particular project. He explains that within a university it is usually easier and more fruitful to cooperate than to compete for resources against colleagues. But he also points out that it is not uncommon for unrelated researchers to converge on ideas and even results at the same time. For example, in 9 December 2009 issue of *Nature*, four separate papers described various features of the receptor for the plant hormone abscissic acid. Stated another way by Abraham Lincoln, "Books serve to show a man that those original thoughts of his aren't very new at all." So, Pinto advises Ginger, that if she wishes to go into an exciting topic, to expect competition, to work hard, be smart, and do the best research possible. And he adds, "Be speedy, Gonzales." Both Ginger and Pinto are excited and, through Ginger's resourcefulness, they find some new collaborators to venture into this new area of research.

During her discussions, a few faculty members at her university tell her about Dr. Carlos Hutten, who is a peatland ecologist in a small applied department at the same university. They suggest she visit him. Dr. Hutten, an older associate professor, has not published extensively, but is a very pleasant person interested in traditional approaches to applied ecology. She makes an appointment to visit with him and during their discussion she notices that he behaves decidedly different than everyone else at the university she'd spoken with about her project. After listening to Ginger excitedly tell of her ideas, Dr. Hutten then describes his own interest as being similar and actually encompassing her own. "I've been a peatland ecologist for 30 years. What does Dr. Pinto know about peatlands? I wrote a proposal 20 years ago about this very topic. I knew global climate change was big while everyone else was still doubters. I even suggested that peatlands would be arbiters and bellwethers of global change." Ginger listened carefully while he continued. "I'm very concerned about you entering this field at this university in Pinto's lab, because you'd be competing against me. In fact, I think Pinto has stolen my research before, which accounts for a good deal of his success. Furthermore, it was because of his theft that my grant proposal wasn't funded." Ginger, now confused, asks what he thinks she should do. She didn't know all these things about Dr. Pinto and she certainly doesn't want to compete with a professor in her own university; especially since, it seems, these were his ideas all along. Dr. Hutten suggests that she really has just two options. One, he explains, is to pick another topic not associated with peatland ecology. "I'm the peatland ecologist," he repeats. The other option is that she works directly with him and not with Dr. Pinto. "I could probably get you a teaching assistantship," he suggests.

After her meeting with Dr. Hutten, Ginger is confused and does not know what to do next. She has suddenly lost excitement for her project and has doubts about continuing in Dr. Pinto's lab. How should she proceed now with all this new information?

1. Discuss the issue with Dr. Pinto and others? Maybe they know more of the history with Drs. Hutten and Pinto. There are at least two sides to every story.
2. Perhaps she should not compete with Dr. Hutten since he was there first? After all, what chance does a PhD student have competing against a professor?
3. Maybe she should simply continue pursuing her ideas for research and surround herself with willing collaborators and competent scientists. After all, Pinto tells her, Hutten does not publish very fast or very often. He tells her of one of Hutten's few graduate students. After the student graduated, it took five years for the paper to get published since Hutten basically sat on it. Pinto had heard other instances where his former graduate students' careers had stagnated because of extreme delays in publishing. While that former student was still the first author, she didn't have much input in the resulting paper. "Don't worry too much about what the competition is doing; worry about controlling your own destiny," Pinto says.
4. It is possible she should switch universities since there are obviously some fishy politics going on where she is currently.
5. Should professors discuss their distaste for one another with students? Should Hutten have criticised Pinto to Ginger? Was it wrong for Pinto to have told Ginger of Hutten's issues with his past students? Where does giving information end and gossip start?

Case study 2: The case of the promising new drug

(Courtesy of Ruth L. Fischbach, PhD, MPE and Trustees of Columbia University in the City of New York; www.ccnmtl.columbia.edu/projects/rcr) (Fischbach and Plaza 2003).

Dr. Linda Roberts has spent the past five years working on a new drug for the treatment of lupus erythematosis. The molecule she designed links a fragment of an anti-inflammatory drug with a protein that binds to the diseased cells. Designing this new drug was made possible by two decades of research in the Immunology Laboratory at Westfield University Medical School, where Dr. Roberts works. Without the basic work in researching the molecular biology of this disease (the early stages of the research were funded by the National Institutes of Health), the highly specific drug would never have been developed.

At the same time that Dr. Roberts' research has yielded such promising results, federal financial support for biomedical research has declined. If this new agent were an effective treatment and a commercial success, it could be extremely helpful for Dr. Roberts, her department, and especially the medical school.

Dr. Roberts' research in the past five years has been supported by funds from Arthrid, Inc., a company that markets a number of drugs for arthritis. She was also given a consultant fee of $50,000. Indeed, researchers working for Arthrid helped with methods for producing large amounts of the therapeutic molecule. Without the resources of a pharmaceutical company, developing a marketable product would have been extremely difficult, if not impossible. Also, changes in federal regulations governing research encourage collaborations between academic scientists and companies in order to promote the transfer of technology from the laboratory bench to the clinic. There has also been a trend for institutions to hold equity interest in the start-up companies of their faculty. Because Arthrid is a local company and has been generous to the medical school, several members of the Westfield University Hospital Institutional Review Board (IRB) have bought stock in the company.

This long-standing relationship made it feasible for Arthrid and the medical school to enter into an agreement that entitles Arthrid to own the patent rights to all discoveries made in the course of the research it funds, and entitles Westfield University to 5% of Arthrid stock and a 5% royalty on sales of all products that result from the research.

In the highly competitive pharmaceutical industry, companies like Arthrid, Inc., seek patents on all promising discoveries. A patent gives the patent holder the right to exclude anyone else from making the patented product during the 20-year life of the patent. During those 20 years, the patent holder would hope to earn enough revenue from the product to recover the typically enormous costs of the basic and clinical research that leads to the production of the product. It is possible for the research and development of a drug, and the subsequent approval by the FDA, to take as long as 10 years and cost more than $800 million.

Included in the Arthrid-Westfield agreement is a non-disclosure agreement, whereby Arthrid will be able to protect its proprietary interests. There are restrictions regarding publication, including Arthrid's right to review all data and a mandate for Dr. Roberts to send to Arthrid all manuscripts at least 30 days prior to their submission for publication. This would allow Arthrid to delete any information that, according to the company's directions, should not be published or presented, which might threaten its rights to any patentable invention.

Extensive use of the experimental drug in animal models of lupus has been highly successful, producing the desired anti-inflammatory effects. Other similar drugs have been used without any serious toxicity. Thus, the drug is now ready for Phase I clinical trials. To encourage this collaboration, Arthrid would be pleased to pay for a trial at Westfield University Hospital. By all accounts, Dr. Roberts would be the ideal clinician to conduct the trial, because of her intimate knowledge of the drug. Arthrid is willing to issue to Dr. Roberts 2% of its common stock. In addition her husband will receive 2% of Arthrid stock and her 14-year-old son will receive 1%. If the trial is successful, this would go a long way toward covering her son's college tuition.

Dr. Roberts and her colleagues work in the hospital's Medical Clinic, which has a large number of patients with lupus who would be available and thus easy to recruit for the clinical trial. Also, because of the department's reputation in research and patient care, Dr. Roberts would be able to enlist the cooperation of other hospital departments around the country in initiating a multicentre clinical trial of the new agent for treating lupus. Dr. Roberts and her colleagues submitted a proposal to the IRB which they believe justifies the use of their clinic patients because of the benefit that this new drug will provide.

On the assumption that she will be conducting the trial, Dr. Roberts approaches her postdoc, Dr. Henry Chung, to ask if he would join her in testing the new drug. But Dr. Chung has been pursuing a different and potentially significant project, cloning a gene for asthma, and is close to completing his work on it. Completing the asthma project would put Dr. Chung in an excellent position to apply for a faculty post and qualify for a grant under a newly announced federal programme; working with Dr. Roberts means that he must set this work aside. Dr. Roberts tells Dr. Chung that if he joins in the Arthrid project, the company will issue him shares of Arthrid common stock equal to 2% and also pay him a generous consulting fee. In addition, Dr. Chung would continue to receive his postdoctoral stipend.

Dr. Chung is already somewhat annoyed that Dr. Roberts is spending so much time at the Arthrid labs; he is not receiving the supervision for his asthma project from Dr. Roberts that he feels he needs. Westfield University

Medical School allows a faculty member to spend 20% of his or her time on outside commitments, and Dr. Roberts is spending about 12–15 hours a week at the Arthrid labs. Since Dr. Roberts works closer to 60 hours a week (she always works on the weekends and takes work home every night), she does not feel that her time away from the medical school is excessive. Furthermore, this time away from her medical-school lab allows her to work on the therapeutic molecule using laboratory equipment at Arthrid which her own lab lacks.

Meanwhile, Dr. Frank Bonita, a colleague of Dr. Roberts', asks Dr. Roberts for a small quantity of a reagent that has been used in the lupus drug research. Drs. Bonita and Roberts were students together, entered the department at the same time, have openly discussed with each other all their research for many years, and are good friends. Each has been indispensable in the research success of the other. The secrecy covenant in the Westfield-Arthrid contract now prevents Dr. Roberts from granting what would otherwise be Dr. Bonita's routine request for a reagent. Dr. Bonita wonders whether Dr. Roberts acted prudently in so restricting herself.

The IRB will meet soon to review Dr. Roberts' proposal to study the new drug at Westfield University Hospital. The IRB chair has been informed by the dean of the medical school how important this proposal is for the medical school and makes the IRB members aware of this.

1. What is a conflict of interest?
2. Why does a conflict of interest matter? Why should the university be concerned?
3. What types of conflicts of interest can you identify in this case?
4. Should Westfield University Hospital undertake the clinical drug trial? If so, should Dr. Roberts participate?
5. Does it matter if Dr. Roberts' financial interest in Arthrid consists of consulting fees, or common stock (equity), or both?
6. Should Dr. Roberts recommend to her patients that they enroll in the clinical trial if it is carried out at Westfield University Hospital? What about elsewhere?
7. Is Dr. Roberts being faithful to her obligation to provide an educational experience for Dr. Chung?
8. Is Dr. Roberts acting properly in the way she chooses to allocate her time? Is this in violation of Westfield University Medical School policy?
9. How should Dr. Roberts respond to Dr. Bonita's request for the reagent?
10. What are the implications of the Arthrid-Westfield non-disclosure agreement for academic freedom?

Judge yourself *redux*

- ✓ Are you susceptible to temptation or greed? Are you especially motivated by money?
- ✓ Do you have diverse interests that could be targets for potential conflicts of interest or commitment?
- ✓ Some people have idealistic delineations that lead them to completely separate science and universities from companies and profit-motives? Do you hold these ideals?

Certainly I like money, but I like to think I'm not greedy. It is better to be paid than not. I tend to not say "no" frequently enough and therefore do have to be concerned about conflicts of interest and commitment. I'd like to think they are all well managed, but sometimes I learn they are not. My ideals are not so idealistic – I'm a pragmatist and a realist. And the reality is I'm happy when I'm productive. Productivity takes many forms as my pursuits diverge from the lab into government and private sectors. I think a lot about managing conflicts.

Judge yourself *redux*

- ✓ When it comes to ideas and projects do you have almost compulsive or obsessive tendencies that could shade your view toward unhealthy competition?
- ✓ What is your style with regards to confrontation and compromise? These are important conflict management tools.

I have been probably too driven before. I'm happy now with putting forth my best thoughts and ideas to test if they are any good at all. I hate confrontation and I'm also not too fond of compromise. But being a scientist requires both to be faced sometimes.

Summary

Conflicts of interest occur when one activity or interest could corrupt or disrupt the pursuit of a constraining interest. Conflicts of commitment occur when opposing commitments in time or energy where one or both will be accomplished suboptimally or not at all. It is important to disclose potential conflicts as they arise and discuss these with university administrators if there is concern. A greater concern for younger scientists is potential conflicts that can occur within the lab and also at the university.

Chapter 13

What Kind of Research Science World Do We Want?

ABOUT THIS CHAPTER

- The best science can be characterised by "a culture of discipline and an ethic of entrepreneurship."
- Without doing science right, i.e., best ethical practices, it is impossible to do the right science.
- The practice of integrity is irreplaceable in research science.

The ultimate goal of science is to gain and apply knowledge to help people, plants, animals, and the environment. Research, both basic and applied, seems to be the best way to gain this knowledge and apply it to address practical problems. How most scientists currently accomplish this goal is to raise funds to allow them to follow their noses in their areas of interest. Funding allows scientists to then perform the right experiments to address testable hypotheses designed to give a decent certainty that the answers obtained are likely the right or wrong ones. In scientific terms, hypotheses can be rejected or they can fail to be rejected. Through the weight of evidence, we can then discern which answers are likely to be the best ones to answer the questions posed. The knowledge is then extensively vetted to the entire world with special interest from other research scientists in the field for verification, criticism, and if appropriate, tacit acceptance. This knowledge is then applied and refined as it then serves as the basis for more research and development by other scientists and engineers. University scientists, perhaps the largest cohort of scientists worldwide, simultaneously perform research while also training the next generation of scientists. A big part of the training is simply allowing students and postdocs to learn by doing under the supervision or (more accurately) apprenticeship by more established scientists. Essentially, this is the science world we have. In many ways, these components are probably optimised for efficiency and success. In practice, of course, there are flaws. There are always flaws when people are involved. Honest mistakes and judgment errors we live with. Science becomes broken when ethical violations occur, especially in terms of FFP, but bad mentoring, in my opinion, is as damaging as some FFP.

Research Ethics for Scientists: A Companion for Students, First Edition. C. Neal Stewart Jr.

Where do we go from here? How can we personally implement improvements in our practices and thereby collectively as the universal group of scientists? Is simple refinement of an already efficient process adequate or do we need to demolish the current system and build a fundamentally different science world from scratch?

"A culture of discipline and an ethic of entrepreneurship"

I would like to suggest that the current system of science has been selected as the most effective one to accomplish what research should do. The system has features that allow science to move quickly while maintaining high quality. A group of smart and disciplined entrepreneurs have essentially selected the system and these are the kinds of people who are currently rewarded and thereby predominantly shape the face of science. They select this system of science and science selects these kinds of people. So, I'd like to end the book on a note of optimism and stress that ideals of science are noble ideals that simply need to be executed with an ethic of integrity. This ethic of integrity combined with noble ideals should result in best practices, many of which were examined in detail in the prior chapters.

In his book *Good to Great*, James Collins (2001) describes common characteristics of companies that were good and went on to become great companies. One of my favourite lines in the book is "When you combine a culture of discipline with an ethic of entrepreneurship, you get the magic alchemy of great performance." I think that this sentence also describes the best scientists who accomplish the best science. You can likely apply this statement to most Nobel Laureates and winners of major science awards – this is how they accomplished great things: self-discipline that yields a disciplined lab setting, which also encourages entrepreneurship. Of course, these men and women of science are not perfect and indeed more than one major prize winner has been implicated in FFP that could certainly have been avoided. But I think it is safe to say that those who have been recognised as the best really are the best. And they inspire us to adopt best practices so our own science can improve.

Ethic of entrepreneurship

Entrepreneurs are people who found their own companies. They have the boldness and confidence to conceive and act on winning ideas that are then brought into fruition to deliver novel and useful products. Replace "product" with "research" and we have now defined the successful academic scientist. University research, by its nature, is a bottom-up enterprise that rewards the successful entrepreneurs who can sustainably find funding for and carry out impactful research. Government and industry science, long known for its top-down management, is becoming more entrepreneurial in nature in which the best ideas and teams are sought in competitive endeavours. There is a convergence on the entrepreneurial model because it is indeed effective in providing sound research solutions to rapidly changing problems. As we recall the philosophical underpinnings of research ethics in Chapter 1, although it was not called the "ethic of entrepreneurship" that is exactly what egoism represents. Recall that egoism is simply defined as people ought to do what is in their own best interest. Entrepreneurs must focus on the narrow set of activities ensuring that their companies or activities will succeed, which, by definition is in their own self interest. One of the surest routes to success is discipline; self-discipline.

Culture of discipline

The best scientists I know are extremely disciplined people. Forget about the stories James Watson tells in his book *The Double Helix* (1968), where he states that he and Francis Crick arrived in the laboratory at 10:00 am and then knocked off at 2:00 pm to have fun. They were intently focused on knowing the structure of DNA and on that scientific question alone. Having some down time to ponder was part of the plan at the time, but in the rest of their careers they were extraordinarily focused and disciplined in all the ways you might imagine scientists being disciplined. That is the only way to have sustained productivity. The best scientists are disciplined to think, read, design experiments, and write grant proposals and papers. The best scientists are disciplined to be competent and sustain competence over the long haul. They are disciplined to treat their trainees well and to give them what they need to, in turn, succeed in science. The best scientists plan their time schedules. They make goals for their days, weeks and years. They have disciplined thoughts and actions. The opposite of discipline is that which leads to ethical problems. Even in the cases of Type A personalities such as we saw in Hwang and Schön, they did not have the patience and discipline to let the scientific process take its course – they made undisciplined shortcuts by fabricating data. It might very well be that the root of most research misconduct is a lack of discipline of one sort of the other. Therefore, the science world I want is one that is described as "a culture of discipline and the ethic of entrepreneurship."

Judge yourself

✓ Are you a disciplined person? Can you channel discipline to produce fruitful science?
✓ Entrepreneurs don't tackle other people's projects but create their own. How entrepreneurial are you?
✓ Do you want to be the best or is being good simply good enough?

Too much pressure?

There exists an argument that the culture of science that focuses on competition, an academic "up or out" tenure system, and "insufficient" research funding is a major cause of problems in research integrity (Munck 1997). I think research integrity is largely a human problem that can be solved, at least partially, by education and the courage to hold people accountable. Eliminating competition and having unlimited funding would not serve to improve the quality of science. I think a lot more available money for research would simply increase the amount of mediocre research and not make the best research better. I posed the following question recently to a group of administrators. If you gave your least productive researchers all the money they wished to have for research, would it elevate their productivity to that of your most productive faculty? No one answered in the affirmative. More funding, while it might be somewhat helpful, will not substitute for competence and discipline in the researcher. But what about the temptation to cheat to win grants? Yes, of course, there are always temptations and the drive for survival. In many ways, there are some practical aspects of the academic system, at least in the US, that might lead to problems.

I think there is immense pressure to push scientists to become one-dimensional workaholics in order to stay competitive in research and to win tenure. This pressure begins in graduate school and tends to peak during the tenure-getting years leading up to the "big decision" whether a university will retain an assistant professor by granting tenure and promotion to associate professor or firing the person. With so much invested in training to get the PhD, doing postdoc stints, and then toiling for years to get tenure, there is a lot at stake for the young scientist who might not earn tenure. I am no fan of the U.S. tenure system. I've now got tenure three times and walked away from tenure twice. I am, however, currently tenured. Certainly, it adds undue pressure in the early years, which might contribute to compromised research integrity, but then tenure protects the deadwood, i.e., unproductive scientists, who do not deserve to be protected. I would rather see a system of five-year contracts in which outstanding scientists would enjoy perpetual five-year contract extensions; low performers might expect experience contract terminations. This would introduce higher accountability while taking a bit of the pressure off in the assistant professor years.

Within-lab competition, covered before, is another topic that can lead to misconduct. A recent case of intra-lab research sabotage, in which a postdoc destroyed the experiments of a graduate student in his lab, is an extreme example of a situation where a scientist internalised pressure the wrong way (Maher 2010). The motivation of sabotage in this case was to make himself appear more productive than a graduate student in the lab. The postdoc was quoted, "I just got jealous of others moving ahead and I wanted to slow them down" (Maher 2010). The science world we don't want is one in which people who can't hold up under a modicum of pressure finds themselves in a very competitive environment, such as in a medical school of a top university.

Another example that has led some people to conclude there is too much pressure in science is the Amy Bishop case. She murdered her department head and two of her colleagues at the University of Alabama-Huntsville in February 2010.

Some blame the pressure cooker of the tenure system, but further analysis of her background shows questionable behaviours and a history of violence even prior to her negative tenure decision. Unfortunately, being denied tenure is the "scarlet letter" of science, branding one as untouchable. Unfortunately, there are few institutions with the prestige to survive dismantling their tenure systems, and then their faculty would probably offer strong resistance. Other institutions would lose their competitive edge to recruit the best faculty if they were to become non-tenure-system islands in a sea of tenure. So, I don't think there will be any big changes in how universities view tenure. I do believe, however, we will see more detailed background checks conducted prior to hiring scientists to avoid another Bishop-like disaster.

The other unchangeable system is that of competitive grants. Biomedical researchers lament decreased funding rates in recent years, but this is because there are probably more biomedical scientists being trained than the funding system can bear. Funding is a product of real needs in various parts of society, politics, and economics. We see politics and societal pressures to land a person on the moon driving the huge budgets on NASA in the 1960s that carried over to the 1970s (see Figure 13.1). We see the oil crisis driving increased research expenditures in bioenergy in the 1980s. This dynamic change of priorities is part of life that scientists should expect and embrace. The most successful scientists love to do research and will adjust their interests to somewhat track future funding cycles and trends. Part of staying competent is embracing change.

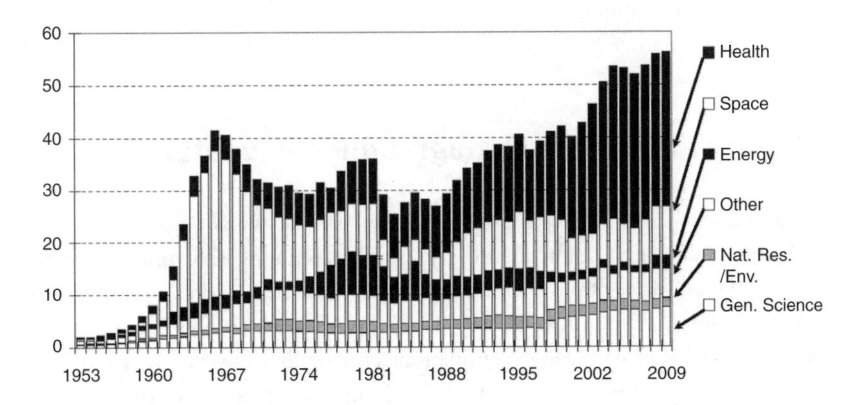

Figure 13.1 Research funding in billions of US dollars in the United States by type, 1953–2009.

Source: Reprinted with permission from AAAS.

As Billy Joel (yes, the musician) observed, "I am, as I've said, merely competent. But in the age of incompetence, that makes me extraordinary." This might be said of many scientists these days. Increasing accountability might also encourage competence. Increasing grant competitiveness in one discipline coupled with more opportunities in other disciplines, should simply shift the areas of science in which people are being trained. History shows us this is how it works. It does no good to bemoan that my favourite area X is no longer being funded, therefore there is something wrong with the system. When we realise the funding system is a social/political/economic chimaera, we should no longer feel the need to make up excuses why the system is flawed. I recall once attending the "research" presentation of a scientist who was on a speaking circuit of sorts, not for his research, but to reassure an audience of like-minded people that funding agencies have lost their missions, etc., which is why nobody in the room was winning as many grants as they did previously. Of course the elephant in the room was that the speaker and much of the audience were simply no longer competitive in the grants world. The truth was that during their careers funding priorities and topics had changed and those scientists, including the speaker, had failed to adapt. Of course, there are other ways to win research funds, such as through companies and foundations. I've enjoyed these too, but I've observed in my own career that competitive grants are one of the primary means by which I remain more competitive and competent in science. If I only relied on earmarks and corporate funding, I would certainly become an undisciplined scientist because exclusively easy money can lull researchers into a lazy stupour. Easy money typically doesn't demand the stringency or rigour in planning research that does competitive research. For that reason I think that a skewed demand and supply spectrum is probably good for science. So, let's take a look at some key features that we can address that will give us the biggest payoff with regards to quality of science research and integrity.

Integrity awareness through ethics education

By far the best thing that established scientists can do for young scientists is to teach ethics in research. Many mentors do just that, but now is the time to institute formalised mandatory ethics training in graduate education at the institution level (Titus and Bosch 2010). I love the one-credit hour research ethics I co-teach with other faculty members. Student evaluations concur that the course makes a big difference in their awareness of ethical issues in research science. In fact, many students wish the course was longer and had more credit hours. A case-study and discussion-based course with 15 students or fewer is rewarding for all who are involved (Stewart and Edwards 2008). In addition, there is also a need for research ethics orientation for postdocs and new faculty as well. In this sort of educational programme, we should focus on the ethical development of the scientist and best practices that give best results.

Accountability

In recent years, scientists have certainly faced increased financial– and results reporting accountability. I suppose that's ok. I think we need another sort of accountability, however. The accountability that I propose is probably better described as joint responsibility for ensuring research integrity. Thus, the focus should be on the science itself and not the scientist. What do I mean? There is a school of thought, usually not overt, that borders on scientist club-ism. What do I mean by that? Simply, that scientists will protect other scientists, especially those in their "club," be it gender, political persuasion, department, commodity, interest, scientific society, or discipline. I think this protectionism and loyalty to friends, instead of that to science or truth itself, is prevalent. It is the enemy of research integrity. Of course, loyalties to people run deep, and there is enduring hope of reciprocal loyalty as well. Therefore, there is a tendency for us to not want to "call out" colleagues, etc. when they err. As we saw earlier there is probably some merit in informal confrontation and rehabilitation within units. One place this informal approach won't work is in the review and handling of manuscripts and grant proposals. We are slow to point out possible ethical missteps, and then editors are sometimes reticent to act on them. I worry about all the new journals being created with the commensurate "need" to publish papers (Siegel and Baveye 2010).

In the urgency to publish, it is not beneficial if both authors and editors cut corners; especially editors. Many journals and scientific societies do not have any ethics statements or procedures on how they will handle breaches of ethics. I have personal knowledge of situations where peer reviewers clearly pointed out plagiarism or research misconduct, only to have an editor not act on them and publish the critically flawed paper anyway. Clearly, editors, especially, must be proactive to maintain "purity" and integrity in science. There is a prevailing view among some scientists that journals exist so that scientists have venues to publish to advance their careers and their students' interests. That is not an accurate view. Journals exist to publish sound science. There is no shortage of self-help/virtual mentor books that are centred on giving advice about how to advance one's career in science. While there is nothing wrong with career advancement per se, that should not be the principle motivation for pursuing a career in science. The main goal should be to advance knowledge. In my opinion, it is in the outlook and motivation where people sometimes go wrong. When we are so set on advancing our own careers and the careers of our pals, we can easily lose sight of research integrity.

We scientists

Incremental science advancements should be rewarded in publications and with grants. Incremental advancement in a specific discipline towards solution of a specific problem is a worthy goal. Yes, breakthroughs are wonderful, but they should not be the objective each morning when the lab door is unlocked. Breakthroughs are often serendipitous, occurring when we least expect them. We need to temper our expectations and feel good about doing sound research that leads to solid advancement.

Now is the time when scientists are becoming proactive about integrity awareness and best practices. A recent "World Congress on Research Integrity" held in Singapore resulted in a statement (www.singaporestatement.org) that lists best practices for research (Ham 2010). With input from delegates from over 50 countries, this statement was written to transcend any nationality or cultural issues with regards to research ethics; i.e., research integrity is for everybody.

We also need to take the time to train the next generation of scientists in best practices; not just how to get ahead or how to win the next grant, but how to truly succeed in science. This is the focus of egoism – a long view for doing what is in one's own self interest need not be self-centred. I like to think that by advancing science and by being flexible, that there will be no shortage of problems for which to apply scientific research. We also need to expect great things from the next generation of scientists. Is that goal at odds with respecting incremental accomplishments? In no way. I have relatively few regrets in my career that has spanned over 20 years, counting my time studying towards the PhD. I regret not being rigorous enough with students, postdocs, and myself. I regret those times when we've taken shortcuts and not engaged in best practices. I regret those times when I've been lazy in not running strong enough when the finish line is in sight. I regret taking such a long time to hone in on research ethics for me and my trainees. From teaching a simple discussion course on ethics in our departments, to being kind and demanding with our students, to being a rigorous reviewer of journal articles: this is the kind of science world that I want. This is the kind of science world that society will respect and reward.

Judge yourself *redux*

✓ Are you a disciplined person? Can you channel discipline to produce fruitful science?
✓ Entrepreneurs don't tackle other people's projects but create their own. How entrepreneurial are you?
✓ Do you want to be the best or is being good simply good enough?

I'm a sporadically disciplined person. I could be better. I do like to think that my group has made worthwhile contributions. I think that entrepreneurship is an asset. I could probably be more aggressive, but I also don't like feeling overwhelmed or that things are out of control. I think that starting a new project and then seeing it until the end is very, very satisfying. The benefits to being the best is that it takes pressure off competing. I'd like to be the best, but many times

I'm content with being on a par with Billy Joel. That is to say, there is room for improvement.

Summary

Science as a way of life is based on discipline and excellence to pursue rigour. Entitlements don't encourage discipline, rigour, or excellence. While pressure in science can be overwhelming to some trainees and scientists, pressure should encourage excellence. Research ethics education is key to becoming better scientists with a culture of integrity.

Appendix

Chapter 4

Case study 1. Choosing between two possible mentors

1. Joel is comfortable with Professor One's hands-on approach, which is more like that of his previous professor. How important is this feeling of comfort and familiarity in making his decision?

   Answer: It is important to feel comfortable and not totally intimidated by your major professor. But at the same time, it is important that MS and PhD students (and postdocs!) be stretched by their scientific training to become competitive and productive. Expanding horizons is probably more important than comfort in science, especially during the student years. Some uphill running is good when you're training for a marathon.

2. How important is it to work on a funded grant or at least, a line of research that seems fundable?

   Answer: Every researcher's goal should be to acquire funding to do publishable work. There are various paths to arrive at that endpoint. Unfunded research might be good for training and for creating a sense of hunger, but given the choice between working on a funded vs. unfunded project the choice is clear: funded.

3. What should Joel think about Professor One's desire to keep the project secret? Do ideas often get stolen in science?

   Answer: There might be good reasons to keep a project secret. If the funder is a company, they might not wish for the public to know all the details from the start. If there is a history of ideas being "stolen" within the university, then we can understand why Professor One is cautious. But on the other hand, in a collegial university environment One's secrecy could be more paranoia than rational. In general students and researchers should be able to freely discuss ideas, projects and science in general with other people at their university and beyond. Too much secrecy in public science is not helpful and indicates that there might be issues with the faculty and/or department.

4. Should Joel worry much about the prospect of working more with Dr. Quattro instead of Professor Two for his PhD? In contrast, he'd be working very closely with Professor One if he chooses his lab.

   Answer: Much depends on the student. Students have different styles of learning, various degrees of confidence and end goals. There are benefits to

Research Ethics for Scientists: A Companion for Students, First Edition. C. Neal Stewart Jr.

working with a younger and more hands-on postdoc compared with a faculty member when conducting research. The postdoc could have more current expertise, more energy for research and could prove to be a valuable mentor in addition to the faculty mentor. In my experience, postdocs can be very beneficial to the training of students.

5. How should Joel feel about the different information he receives about the supercomputer? How should that affect his decision?

   Answer: The differing opinions from the faculty members about the necessity and accessibility of the supercomputer forces the student to root out the information personally and shows how important it is to have an understanding of the field of science. Discussing this issue with other people in the lab and other faculties in the field as well as being familiar with the literature are absolutely essential to discover the truth. Many a lie is told during recruiting by the recruiter and students need to be less gullible.

6. How should Joel feel about One's guarantee that Joel would complete his PhD in three years versus the absence of any guarantees from Two?

   Answer: The completion of a PhD is much less predictable than that of masters or bachelors degrees since students must essentially prove their independence in science. Independence comes at various speeds among students. Also, any science worth doing could fail and failed experiments will affect timelines. Clearly if Professor One is guaranteeing a degree in three years then either the training is formulaic and therefore not producing independent thinking and innovative research or One is misleading Joel.

7. Which lab should Joel join and why?

   Answer: This is not a question of a small lab versus a big lab. There are components of integrity with regards to secrecy, time toward degree and perhaps also questions of honest competence with regards to needing a supercomputer that should cause a student to probably disqualify Professor One as a potential mentor. This doesn't mean the student should necessarily choose Professor Two. There are always intangibles that come into play when choosing a mentor, but students need to go into the process being well read and with their eyes open. Students who know what they want and where they want to go have a better chance at experiencing a fulfilling studentship than those students who are at the whim of a faculty member.

Chapter 4

Case study 2. Choosing the right research project: the new graduate student's dilemma

1. What factors should Karen weigh up when making her final decision for her graduate programme?

Answer: Since she is a new graduate student just beginning a life of research, she has many questions with probably fewer answers. She should think about her end goals and passions and put those in the context of practical matters. She should think about timelines, how much energy and dedication she would have for research in general and each topic, in particular. She should think about how important creativity is to her vs. knowing that a project will proceed on a more certain timeline. Given her inexperience, there is a fair amount of uncertainty about either decision. One could prove to eventually be much better than the other. Certainly, taking the molecular project has higher risks and greater potential rewards. Much of the decision has to do with how much inherent risk Karen is willing to assume at such an early stage in her training.

2. For the toxicology project she will be a more independent worker, but not play a creative role in experimental methods, since those have been finalised in the grant proposal. How much should that matter to Karen?

   Answer: Again, much depends on Karen. The best scientists are very creative people and without the freedom to be creative, they would lose interest in a project. If Karen were to take the molecular project, her creativity would likely be challenged to some degree by the more experienced postdoc on the project. And so if she were to take the molecular project, she would have to be certain she is compatible with the postdoc.

3. A tradition doctorate programme in biology ranges from three to five years, depending whether a masters degree is earned, how fast the research becomes productive, and other factors. If Karen pursues the doctorate with the molecular study she is not guaranteed funding after the third year. How important is not having guaranteed funding after three years to the student?

   Answer: Karen has more flexibility than she probably knows at this point. If she were to take the toxicology project with its sure funding for two years, even if it takes her more than two years to finish the project her stipend funding would probably not be in jeopardy. Then after she completed the masters, she would have the flexibility to move to another lab or university if she so wishes – at least she would be in a good situation to move. If she took the molecular project, she is probably looking at five years, at least, to complete her PhD. If she catches on to the techniques quickly, is industrious and has a knack for research, she could conceivably play an important role in landing another grant in this area that would help her to finish her degree. Dr. Leavens certainly expresses confidence in the fundability of the molecular research. Here is a secret. Professors will find a way to keep the best students funded. So while students should be aware of funding constraints and timelines, they should work to be the very best researchers and increase their value. If Karen does take the molecular project and funding ends before she can finish in five years – say the funding ends in three years, but Karen has experienced at least some success – she can probably take the masters degree and move on to another lab if needed. I'd take the riskier but more fulfilling project if I were Karen. (But during the first time this case study was discussed in the ethics class – this particular section having mostly new masters students – the members voted 5 to 1 in favour of the applied toxicology study.

Chapter 5

Case study. The case of the missing mentor

1. Should Mitch try a different strategy to get good mentoring from his major professor? If you think this is a good solution, how might he approach it? Is Mitch blameless in his situation?

   Answer: Clearly the present strategy is ineffective. Mitch is not blameless in that he is allowing bad mentoring to dominate his graduate programme and apparently Dr. DeBague is unaware that this is even happening. Perhaps Mitch could invite his mentor to meet him in the laboratory; perhaps to discuss a certain piece of equipment or method. Perhaps Mitch could request meetings at strategic times of the day, say before lunch or at the end of the day, where there could be some time constraints. Perhaps Mitch could have a written agenda of items to discuss that might facilitate more productive meetings. It might be that all these approaches and a few others are needed to break the destructive cycle of non-productive meetings with his mentor.

2. What is the chance of success should Mitch decide to continue to go it alone as he has in hopes that Dr. DeBague is right? Do you believe the assertion that Mitch will be successful if he works hard? After all, isn't developing scientific independence a key outcome during the PhD process? What are the advantages and disadvantages of taking this approach? Is Mitch working towards independence?

   Answer: Clearly the main reasons that Mitch is in Dr. DeBague's lab is because of the match of expertise and interests, plus a positive recommendation from his trusted masters degree mentor. Even though the recommendation is clearly out-of-date, the area of expertise is still relevant. If Mitch does stay in the DeBague lab, he will likely have a lonely research existence and he must ask himself if he has the personality, fortitude and independence to prosper in such a situation. Presently Mitch's rate of becoming scientifically independent is probably too slow. Mitch can get technical help and mentorship from outside his lab or even his university. Trainees should, in most cases, avoid too much insularity and isolation.

3. Should Mitch leave the university or simply leave his department? Or should he stay in his department but switch labs? Or something else? He's taken a class with another professor who is engaged in research that is different from that which Mitch originally wanted to do, but Mitch has talked extensively about his problems with the other professor's graduate students and has seen merits in the new area; he thinks he might be able to cultivate interests in this area. The other students confide that since it is a larger lab they aren't alone in science,

but also complain that their professor is also very busy. However, they say they do see him and get directions and advice, but without the travelogue. Their professor is also responsive to email and reads and comments on their written projects – it only takes a few days to get back constructive comments. They encourage him to speak to their professor about switching labs. What are the advantages and disadvantages to switching labs and major professors? Would it place the alternative professor in a collegial bind to accept Mitch? Should Mitch change departments or universities to salvage his science career without Dr. DeBague?

Answer: It will be the rare graduate student who will truly prosper in the current DeBague lab, which might account for its very small size. Politically, it is very difficult to remain within the same department and switch professors because of collegiality issues; e.g., "you stole my graduate student." A better solution is to leave the current university and find another mentor who is in the field that Mitch is passionate about. If Mitch were to go to the professor's lab that his friends recommend in the same department, the only conditions for which that will be workable are: 1) Mitch sincerely has a change of heart in research interests. This gives him a legitimate way out; 2) Mitch discusses this change of heart with Dr. DeBague and seeks an amicable release. This could still result in a stressed situation, however. The optimal situation is to train in the best lab with the best mentor in the student's area of interest.

4. Are there any other potential solutions?

   Answer: Maybe Mitch should leave school and work as a researcher for some time. Under very, very few circumstances would I recommend any student to remain in the DeBague lab, which appears to be fatally toxic.

Chapter 8

Case study 1. Who is an author?

1. Why should Ms. Jacobs and Dr. Frank have discussed the laboratory's approach to authorship issues when she started working in his laboratory?

   Answer: The PI typically sets lab policy, including authorship. There was a sizable disconnect in philosophy between Dr. Frank and Ms. Jacobs and the student got surprised much later when she learned how far apart they were. If either she or he had broached the subject, there probably would have been no misunderstandings when the actual papers were written.

2. Why is the order of authorship and the listing of authors important in a research paper?

   Answer: There are customs in various fields of science with regards to authorship order and readers infer the roles of various people based upon position. One system to disambiguate roles and order is one that requires authors to list contributions.

3. What is the difference between an acknowledgment and a listing as an author?

   Answer: Being an author means you made significant contributions on a paper whereas being listed in the acknowledgments means you simply helped in a study. Acknowledgments are not listed on CVs whereas your authored papers are. Therefore, scientists should be motivated to significantly contribute to studies to earn authorship.

4. Although many journals subscribe to the guidelines of the International Committee of Medical Journal Editors, many do not, and many researchers do not follow the practices that it recommends. What tends to happen, and how are ICMJE standards being challenged?

   Answer: Universal standards are good for people who enjoy universal standards. Many scientists are "free agents" with strong opinions on how things should be done. There are also ingrained customs, traditions, and special circumstances that conspire against any set of standards.

5. Who among the authors takes responsibility for submitting the paper to a journal and following up with the editor and peer-review revisions?

   Answer: The corresponding author agreed upon by the set of authors is responsible for these duties. It is typically the PI, but does not have to be.

6. What are some potential problems with Dr. Frank's submitting a paper on preliminary findings and not performing sufficient corroboratory experiments?

   Answer: The science could be wrong, and reviewers and editors will also tend to recognise the preliminary nature of the work.

7. What kind of problems may arise if the same data is used in multiple papers in the research literature?

   Answer: This is a serious ethical problem that undermines the assumption that each peer-reviewed journal article contains novel data and contributions.

8. What might happen if someone is listed as an author on a paper for which he or she did not do any work?

   Answer: First, the other authors could be resentful, leading to an unhappy lab. The PI that allows this to happen is undermining his or her programme. Second, it is an ethical mistake to ascribe authorship to a person who does not deserve it in that it is deceptive. Third, if there is a problem with the paper, there will be the expectation for all authors to assume at least some level of responsibility, a task that is not reasonable for someone not involved with the study.

9. What might have been done to resolve Ms. Jacobs's ethical dilemma with Dr. Frank about the authors on the paper?

   Answer: Since Dr. Frank is the PI and since he has already created an expectation of authorship among many lab members he added on as authors, it would be difficult to fall back to Ms. Jacob's position to implement ICMJE standards on this paper. She could still strive, however, to request the new authors to make significant contributions to the paper. She is in a delicate situation. She could try to submit the paper to a journal in which author contributions are required to be listed in order to disclose actual contributions.

Chapter 8

Case study 2. The case of the ghost-writing student

1. What is the best solution for Smyth to continue his current research and graduate?

   Answer: Smyth is not in a good situation. He finds himself with an unethical professor in a field that has a difficult funding landscape, and is someone who associates with unethical people, in this case Johnston, the part-time student and now full-time administrator who is bribing the professor with funding. Sometimes the best course to take is cut your losses and start over.

2. "Ghostwriting" is defined as writing in the name of another. There are many commercially operated "ghostwriting" websites. Is it ethically acceptable to publish ghostwritten scientific papers?

   Answer: A ghostwritten business document or autobiography is in a different ethical world than ghostwritten science (unethical). This is because there are technical, expert, and professional expectations that scientists do their own writing, which is not the case in some other fields. There is also a level of responsibility expected for authors of scientific papers that doesn't exist in some other fields where ghostwriting is acceptable.

3. There are some science fields that acquiesce "ghostwriting." Some people argue that writing is not a part of sciences, thus ghostwriting is acceptable unlike data fabrication or stealing intellectual property. They also say that you can be more productive if you are working in the lab instead of spending a lot of time to writing papers. What do you think about this opinion?

   Answer: Ghostwriting is becoming more prevalent in the biomedical sciences, for instance. While it is probably not on the same level of offense as FFP, it is probably sanctionable and is not a good practice in research. Scientists need to be careful about authorship and ethical concerns in all documents with which they have interaction and participation. This non-publication example was offered to show that authorship issues extend beyond primary research papers. Scientists should approach authorship with the same degree of care and passion as their experiments. Reputations have been made and broken with authorship of key scientific publications.

4. Are there any other ethical problems in this case study?

   Many. See above.

Chapter 9

Case study 1. The case of a questionable grant proposal

1. Funding is one of the most important elements in scientific research. Writing grant proposals is one of the major tasks for most PIs. Does Dr. Kohls do anything wrong by asking his postdoc to write most of a grant proposal but with Dr. Kohls as PI? Is it a misuse of NSF funds to ask the postdoc to write the follow-up grant?

   Answer: To answer the second part of the question first, this could go either way. When Kohls requested funds from the NSF to train a postdoc, he probably said that the postdoc would be working 100% of the time on the project. However, proposal writing and other professional activities are considered to be valuable training opportunities. If the postdoc stopped working on the NSF-funded research to focus solely on the proposal-writing, then that would be inappropriate. Researchers must learn to juggle multiple responsibilities while having auditable fidelity to the funding agent. Care must be taken not to dilute effort promised to a funded project. To answer the first part of the question, no, Dr. Kohls did nothing wrong to ask the postdoc to author the grant proposal. This is common and very good practice, which provides a valuable training opportunity for the postdoc to learn how to write successful proposals. Again, care should be made not to abuse the postdoc; such a request must be entirely voluntary and on top of the postdoc's normal duties.

2. Grant proposals are not peer-review publications. Is it ok to be less restrictive for plagiarism and intellectual property issue in writing grant proposals compared to peer-review publications?

   Answer: No! You should not steal ideas, data, phrases, sentences that are not your own from any source.

3. You have access to someone's idea by reviewing a submitted manuscript. You review the manuscript and reject it. However, you are interested in the concept or experimental designs in the rejected manuscript. The rejected manuscript has never been published later in any journals, thus you can't cite a reference. What is the best way to utilise the idea or experimental designs in your grant proposal?

   Answer: There is no ethical way to use the idea without the original source's permission. While a delicate line to walk, it might be possible to engage the originator and attempt to make this person a collaborator. Many people would be flattered to be contacted about their novel ideas and could also be seeking a collaborator. One must keep in mind that the person will probably not be happy about the rejected paper however. Typically anonymous reviewers do not contact authors of the papers they review; it is open to discussion whether there would be ethical prohibitions from doing so. However, peer reviewers can waive anonymity when reviewing submitted manuscripts, which might make future contact less of a surprise.

4. Was it wrong to give the postdoc the rejected journal submission?

   Answer: Yes. If the intent had been to educate the postdoc on how not to write a paper or to request help in reviewing it, then most people believe it is allowable to let a labmate read a confidential submission. In this case the submitted paper must still remain in confidence to the postdoc; i.e., it should not be passed on further. However, it is ethically wrong to give the postdoc the paper with the intent of stealing and using the idea presented in the confidential document.

5. Was it wrong to penalise the postdoc for conscientiously objecting to participate in the proposal writing and seeking another position?

   Answer: It is wrong to penalise the postdoc. Indeed, it is stupid to penalise the postdoc since the postdoc has solid grounds to report Kohls to the NSF and his institution for falsification after the grant proposal was submitted. People leave labs for all kinds of reasons and it is best practice to be generous and help trainees to advance in their careers.

Chapter 9

Case study 2. The case of the collaboration that couldn't

1. What are the ramifications of including a potential co-PI's data after the collaboration is dissolved?

   Answer: It is never allowable. Wren sent Pfaff the data to be used in the proposal that Wren thought would be submitted. With the partnership called off, the data should not be used. Even if they had teamed up successfully on the proposal, it would have been a further breach of etiquette for Pfaff to use Wren's data for any other purpose. Wren's data, text and ideas should have been redacted from the proposal prior to its submission.

2. Why did Wren decide to enter, and then resign from the collaboration? Was he at fault? What drove his decisions?

   Answer: Institutional pressures can be intense. It is certainly poor form to agree to partner and then renege, especially at very short notice. He entered the collaboration based partly on feelings and partly on the need and desire to collaborate. Wren probably thought this collaboration could expand his reach in research. He became convinced from his colleague (boss) that being on two potentially competing proposals could put their own centre proposal in jeopardy. Whether it would actually jeopardise it is debatable, but again, institutional pressures and politics can be powerful forces. It probably would have been better if Wren had discussed the idea of the collaboration with his centre director and others before agreeing to collaborate with Pfaff.

3. Do you think there could be an actual conflict of interest if Wren were to be part of his own institution's (UB) centre and RAM IT's mini-centre?

Answer: Probably not. It might have set up perceived conflicts and also set up situations in which Wren would have had to make some sensitive decisions about what data and ideas to share since the two centres would have been in the same area.

Chapter 10

Case study. What is responsible peer review?

1. What types of conflict of interest might arise when someone is asked to review a paper or grant application?

   Answer: The case study illustrates several quandaries that arise during peer review. Authors of submissions really want the best and most qualified reviewers – those who are most closely aligned with research in question. Experts in the field, however, are also in the position to best take advantage of privileged information in the wrong way, which was illustrated here. Sometimes authors request that scientists that are very very closely aligned, i.e., competitors, not be assigned to review their papers. This is often completely appropriate. More obvious conflicts of interest arise when colleagues, friends, relatives, former lab mates, collaborators or enemies are asked to review papers. There can be a temptation to focus too much on the authors and not enough on the science. There could exist desires to unduly help the authors by giving a soft review or unduly harm authors for punitive and personal reasons. Editors tend to minimise conflicts of interest where possible. It is perfectly appropriate for reviewers to declare potential conflicts of interest to editors when chosen to review a paper of concern.

2. Is it ever appropriate for a peer reviewer to give a paper to a graduate student for review? If so, how should the reviewer do so?

   Answer: It is appropriate to give a paper to a trainee as part of training. It is often useful to authors and the journal as additional light can be shed on a submission. The paper should always be held in highest confidence and data or ideas should never be used for unethical purposes. These opportunities should be taken to instruct trainees on best ethical practices. The peer reviewer should also review the paper and meld the reviews between the assignee and student. It is inappropriate in most cases for graduate students to be stand-alone reviewers since they are considered to be scientists-in-training and not yet qualified to be sole reviewers. The submission and notes should be destroyed after the review is complete.

3. Is it appropriate for a peer reviewer to use ideas from an article under review to stop unfruitful research in the reviewer's laboratory?

   Answer: The obvious answer is "no" since the information is privileged. But desperate researchers do many things to stem the tide of unproductivity. If one takes the perspective of what's best for science, then it could be entirely appropriate to stop bad experiments on any source of information. However, it is a slippery slope to now use privileged information to start new experiments – see next question.

4. Is it ever appropriate for a reviewer to use ideas from a paper under review, even if the reviewer's method to achieve a result is different from that used in the paper under review? If so, how should the reviewer proceed?

   Answer: The purpose of the peer-review is to help the journal editor make a decision and to help the authors improve their papers. The purpose of peer review is not to help the peer reviewer. However, peer reviewers are motivated to review papers to become more knowledgeable, so the process is not entirely altruistic. It is important that reviewers resist the temptation to cross the line and do something with privileged information that reviewers would not want done to them (Golden Rule). There are, at least, two options, however, to use the privileged information in a way that does not compromise the integrity of peer review. First, the reviewer could simply wait for the paper to be published and then use the information garnered during the peer review. There is a time lag, but often papers are published soon after review is completed – weeks to months. Of course this route is predicated on a positive decision of the editor to accept the submission. A second option, not done very often I suspect, is to perform a non-anonymous peer review so that the reviewer is then known to the author. Perhaps a collaboration can be subsequently established. Editors typically do not want reviewers to directly contact authors with their reviews. But I don't see anything wrong with reviewers contacting authors after the peer review is communicated by the editor to authors.

5. What are some of the challenges in the current peer-review process, in which the peer reviewer is anonymous but the author is known to the reviewer?

   Answer: One obvious drawback is that bias can creep into reviews when reviewers know the identity of authors. Some journals now redact the author names on review copies. However, even with names redacted, it is typically not difficult to figure out who the authors are if they are in the same field as the reviewer.

6. What recourse is there for Dr. Morris if he suspects that his ideas were plagiarised?

   Answer: He can notify the editors of *Protein Interactions* and *Science* simultaneously about the suspected plagiarism. In addition, he could notify funding agencies and the research officer of Dr. Leonard's university. I think that most investigators in the same position as Dr. Morris would do neither, and very rarely contact funders and institutions. If I felt wronged by such an action, and I felt strongly, there could be the thought that I should push for retraction of the *Science* paper. The problem is that it might be difficult to prove wrong doing and many scientists don't relish battles of this sort. However, let's say that Dr. Morris does want to fire a shot over the bow, but does not want to press for official action. He could confront Dr. Leonard and/or the graduate student about his suspicions. It could act as a deterrent in the future. It could scare the graduate student into not being a repeat offender. I think most scientists will simply complain to friends and not take any assertive actions.

Chapter 12

Case study 1. The case of the crowded room – who should investigate a line of research when many people have an interest in it?

Overriding question: In context of the situation, what should Ginger do next?

1. Discuss the issue with Dr. Pinto and others? Maybe they know more of the history with Drs. Hutten and Pinto. There are at least two sides to every story.
   Answer: This is a good idea. She obviously trusted Dr. Pinto enough to join his lab. Her mentor should be the main person from whom she seeks advice. Other students might also be aware of interactions, office politics and the history among different professors; they might have helpful information.

2. Perhaps she should not compete with Dr. Hutten since he was there first? After all, what chance does a PhD student have competing against a professor?
   Answer: Maybe, but maybe not compete. Dr. Hutten does not publish very fast, so there might be a chance that the project is not performed if she doesn't do it. If Ginger does pursue the project she envisages, there could be some conflict, especially in the early stages of the project. So, much of the answer depends on how Ginger views conflict in general. The story might be different if Pinto and Hutten were collaborators and friends, where conflict would probably need to be minimised, but there would seem to be no love lost no matter what her decision.

3. Maybe she should simply continue pursuing her ideas for research and surround herself with willing collaborators and competent scientists. After all, Pinto tells her, Hutten does not publish very fast or very often. He tells her of one of Hutten's few graduate students. After the student graduated, it took five years for the paper to get published since Hutten basically sat on it. Pinto had heard other instances where his former graduate students' careers had stagnated because of extreme delays in publishing. While that former student was still the first author, she didn't have much input in the resulting paper. "Don't worry too much about what the competition is doing; worry about controlling your own destiny," Pinto says.
   Answer: This is probably very good advice.

4. It is possible she should switch universities since there are obviously some fishy politics going on where she is currently.
   Answer: There doesn't seem to be sufficient justification to warrant this decision. All organisations, and certainly all research universities have their

own particular politics that must be acknowledged. Every place has a group of professors, justly or unjustly, who believe that every good idea they ever had was stolen and someone else received the credit that they themselves are due. Professors that have special funding, such as that assured by HHMI, are often viewed as having an unfair advantage and entitlement compared with average faculty members. On the outside chance that Pinto did indeed steal Hutten's idea, it is highly unlikely that all Pinto's success came from idea plagiarism. That much of his success came from practicing FFP is not impossible, but it is unlikely. After so many years, one would think that a habitual FFPer would have been caught.

5. Should professors discuss their distaste for one another with students? Should Hutten have criticised Pinto to Ginger? Was it wrong for Pinto to have told Ginger of Hutten's issues with his past students? Where does giving information end and gossip start?

   Answer: When there is a long history of animosity and distrust, the normal line of etiquette is sometimes blurred. This is especially the case since Hutten essentially disparaged Pinto to his graduate student and essentially tried to "steal" her. Hutten likely crossed the line of fair play, and perhaps so did Pinto. Gossip is spreading false rumours and often of a personal nature. At least the professors have not engaged in that activity, which would be in poor taste.

Chapter 12

Case study 2. The case of the promising new drug.

1. What is a conflict of interest?

   Answer: Conflicts of interest occur when one activity or interest could corrupt or disrupt the pursuit of a constraining interest.

2. Why does a conflict of interest matter? Why should the university be concerned?

   Answer: This case study demonstrates the depths to which conflicts of interest can impact a faculty member, research programme, and university. The university is deep into a relationship that has probably crossed the line in propriety, at least in some aspects.

3. What types of conflicts of interest can you identify in this case?

   Answer. Dr. Roberts is taking too much consulting time. Most universities count days per week doing consulting when making the calculation, e.g., 20% consulting would result in being absent from university duties and doing consulting one day per week, or 8 hours per week. It matters little about typical numbers of hours she works each week. She has a conflict of commitment evidenced by not adequately training her postdoc. Her shares and her family's shares of stocks upon a successful clinical trial is a conflict of interest, as is the situation with the hospital IRB. There are probably others to list.

4. Should Westfield University Hospital undertake the clinical drug trial? If so, should Dr. Roberts participate?

   Answer: It would not be the best practice. If Westfield did decide to undertake the clinical trial IRB members would need to be divested of stock and Dr. Roberts should not participate.

5. Does it matter if Dr. Roberts' financial interest in Arthrid consists of consulting fees, or common stock (equity), or both?

   Answer: Appropriate amount of consulting is fine, and so is stock if there is no connection between her performance or the studies performance and the stock. She should never be placed in a position to affect stock price and financially benefit with her science.

6. Should Dr. Roberts recommend to her patients that they enroll in the clinical trial if it is carried out at Westfield University Hospital? What about elsewhere?

   Answer: No and no.

7. Is Dr. Roberts being faithful to her obligation to provide an educational experience for Dr. Chung?

   Answer: No. She is not only shortcutting him with her energy and focus, she is trying to distract him with opportunities and money that are not in his long-term interest.

8. Is Dr. Roberts acting properly in the way she chooses to allocate her time? Is this in violation of Westfield University Medical School policy?

   Answer: She is consulting too much – see the answer to Question 3.

9. How should Dr. Roberts respond to Dr. Bonita's request for the reagent?

   Answer: Unfortunately, she must explain to Dr. Bonita that her contractual agreement disallows sharing of the proprietary reagent.

10. What are the implications of the Arthrid-Westfield non-disclosure agreement for academic freedom?

    Answer: As long as she ultimately has the right to timely publication of the data, there is no conflict.

References

Akst, J. 2010. I hate your paper. *The Scientist* **24**(8):36–41.

Allen, J. 2008. Can of worms. Retrieved 2009, from *On Wisconsin*, University of Wisconsin Alumni Association: http://www.uwalumni.com/home/alumniandfriends/onwisconsin/owspring2008/worms.aspx.

Anderson, M. S., Horn, A. S., Risbey, K. R., Ronning, E. A., De Vries, R. and Martinson, B. C. 2007. What do mentoring and training in the responsible conduct of research have to do with scientists' misbehaviour? Findings from a national survey of NIH-funded scientists. *Academic Medicine* **82**(9):853–860.

Angell, M. 2001. Medicine in the noise age: what can we believe? *Accountability in Research* 8:189–195.

Anonymous. 2005. Timeline of a controversy. Retrieved 2009 from *Nature News:* http://www.nature.com/news/2005/051219/full/news051219–3.html

Anonymous. 2006. Beautification and fraud. *Nature Cell Biology* 8:101–102.

Anonymous. 2009. Editorial: Credit where credit is overdue. *Nature Biotechnology* 27:579.

Anonymous. 2010. The sequence explosion. *Nature* 464:670–671.

Ashworth, P. and Bannister, P. 1997. Guilty in whose eyes? University students' perception of cheating and plagiarism in academic work and assessment. *Studies in Higher Education* 22:187–104.

Associated Press. 2009. Senator's affair highlights capitol temptations. *Knoxville News Sentinel* August 10, 2009, p. A4.

Augenbraun, E. 2008. Letter to the editor. *The Scientist* **22**(5):17.

Barnbaum, D. R. and M. Byron. 2001. *Research Ethics: Text and Readings.* Prentice Hall, Upper Saddle River, NJ.

Bauerlein, M., Gad-el-Hak, M., Grody, W., McKelvey, B. and Trimble, S. W. 2010. We must stop the avalanche of low-quality research. *The Chronicle of Higher Education.* Published Online: http://chronicle.com/article/We-Must-Stop-the-Avalanche-of/65890/.

Berlin, L. 2009. Plagiarism, salami slicing, and Labachevsky. *Skeletal Radiology* 38:1–4.

Bordons, M. and I. Gomez. 2000. Collaboration networks in science. Pp197–213 in *The Web of Knowledge*, Cronin, B. and H. B. Atkins (eds.). American Society for Information Science, Medford, NJ.

Borrell, B. 2010. A pioneer's perils. *The Scientist* **24**(11):19.

Breen, K. J. 2003. Misconduct in medical research: whose responsibility? *Internal Medicine Journal* 33:186–191.

Breitling, L. P. 2005. Misconduct: pressure to achieve corrodes ideals. *Nature* 436:626.

Research Ethics for Scientists: A Companion for Students, First Edition. C. Neal Stewart Jr.

Brown, A. S. and Murphy, D. R. 1989. Crytomnesia: delineating inadvertent plagiarism. *Journal of Experimental Psychology: Learning, Memory, and Cognition*. 14:432–442.

Brumfiel, G. 2008. Physicists all aflutter about data photographed at conference. *Nature* 455:7.

Brumfiel, G. 2009. Breaking the convention? *Nature* 459:1050–1051.

Butler, D. 2010. Journals step up plagiarism policing. *Nature* 466:167.

Carlson, S. 2010. Lab notebook tips from a patent litigator. *Genetic Engineering News* **30**(1):10–11.

Carroll, J. and Appleston, J. 2001. *Plagiarism: A Good Practice Guide*. Oxford Brookes University, UK.

Carroll, J. 2007. *A Handbook for Deterring Plagiarism in Higher Education*. Second Edition. Oxford Centre for Staff and Learning Development, Oxford Brookes University, UK.

Chalmers, I. 1990. Underreporting research is scientific misconduct. *The Journal of the American Medical Association* **263**(10):1405–1408.

Collberg, C., Kobourov, S., Louie, J. and Slattery, T. 2003. Splat: A System for Self-Plagiarism Detection. Retrieved 2010, from the Department of Computer Science, University of Arizona: http://splat.cs.arizona.edu/icwi_plag.pdf.

Collberg, C. and Kobourov, S. 2005. Self-plagiarism in computer science. *Communications of the ACM* **48**(4):88–94.

Collins, J. 2001. *Good to Great*. Harper Business, New York.

Comstock, G. L. (Ed.) 2002. *Life Science Ethics*. Iowa State Press, Ames, IA.

Couzin, J. 2006a. U.S. rules on accounting for grants amounts to more than a hill of beans. *Science* 311:168–169.

Couzin, J. 2006b. Stem cells . . . and how the problems eluded peer reviewers and editors. *Science* 311:23–24.

Couzin, J. 2006c. Truth and consequences. *Science* 313:1222–1226.

Couzin-Frankel, J. 2010. The legacy plan. *Science* 329:135–137.

Croll, R. P. 1984. The non-contributing author: an issue of credit and responsibility. *Perspectives in Biology and Medicine* 27:401–407.

Cyranosaki, D. 2006. Rise and fall. Retrieved 2010 from *Nature News*: http://www.nature.com/news/2006/060111/full/news060109–8.html

Dingell, J. D. 1993. Misconduct in medical research. *The New England Journal of Medicine* 328:1610–1615.

Dreyfuss, R. 2000. Collaborative research: conflicts on authorship, ownership, and accountability. *Vanderbilt Lab Review* 53:1159–1232.

Eastwood, S., Derish, P., Leash, E. and Ordway, S. 1996. Ethical issues in Biomedical Research: perceptions and practices of postdoctoral research fellows responding to a survey. *Science and Engineering Ethics* 2:89–114.

Errami, M. and H. Garner. 2008. A tale of two citations. *Nature* 451:397–399.

Fanelli, D. 2009. How many scientists fabricate and falsify research? A systematic review and meta-analysis of survey data. *PLoS ONE* 4:e5738.

Fischbach, R. L. and Plaza, J. 2003. The case of the promising new drug. Retrieved 2009, from Responsible Conduct of Research, Columbia University: http://ccnmtl.columbia.edu/projects/rcr/rcr_conflicts/case/index.html.

Fowler, A. and R. E. Kissell, Jr. 2007. Winter relative abundance and habitat associations of swamp rabbits in eastern Arkansas, *Southeastern Naturalist* 6:247–258.

Gert, B. 1997. Morality and scientific research, pp. 20–33. Published in Elliott, D., and Stern J. E. (eds), *Research Ethics*. University Press of New England, Hanover, N.H.

Green, L. 2005. Reviewing the scourge of self-plagiarism. Retrieved 2010 from *M/C Journal* **8**(5). http://journal.media-culture.org.au/0510/07-green.php.

Gunsalus, C. K. 1998. How to blow the whistle and still have a career afterwards. *Science and Engineering Ethics* 4:51–64.

Ham, B. 2010. Singapore statement urges global consensus on research integrity. *Science* 329:1615.

Handwerker, H. 2010. Impact factor blues. *European Journal of Pain* 14:3–4.

Hard, S. F., Conway, J. M. and Moran, A. C. 2006. Faculty and college student beliefs about the frequency of student academic misconduct. *The Journal of Higher Education* 77:1058–1080.

Harris, R. 2001. *The Plagiarism Handbook: Strategies for Preventing, Dectecting, and Dealing with Plagiarism*. Pyrczak Publishing, Los Angeles, CA.

Hayden, E. C. 2008. Chemistry: designer debacle. *Nature* 453:275–278.

Heim, R. and Tsien, R.Y. 1996. Engineering green fluorescent protein for improved brightness, longer wavelengths and fluorescence resonance energy transfer. *Current Biology* 6:178–182.

Herman, I. P. 2007. Following the law. *Nature* 445:228.

Hirsch, J. E. 2005. An index to quantify an individual's research output. *Proceedings of the National Academy of Sciences USA* 102:16569–16572.

Howard, R. M. 2001. Forget about policing plagiarism. Just teach. *The Chronicle of Higher Education*. Published Online: http://chronicle.com/article/Forget-About-Policing/2792/

Institute of Medicine (U.S.) Committee on Assessing Integrity in Research Environment, National Research Council, and Office of Research Integrity. 2002. *Integrity in Scientific Research: Creating an Environment that Promotes Responsible Conduct*. The National Academies Press, Washington, DC.

Judson, H. F. 2004. *The Great Betrayal: Fraud in Science*. Harcourt Inc., Orlando, FL.

Kearns, H. and Gardiner, M. L. 2011. The care and maintenance of your advisor. *Nature* 469:570.

Koocher, G. and Keith-Spiegel, P. 2010. Peers nip misconduct in the bud. *Nature* 466:438–440.

LaFollette, M. C. 1992. *Stealing into Print Fraud: Plagiarism, and Misconduct in Scientific Publishing*. University of California Press, CA.

Lau, J., Ioannidis, J. P. A., Terrin, N., Schmid, C. H. and Olkin, I. 2006. The case of the misleading funnel plot. *The British Medical Journal* 333:597–600.

Levine, R. J. 1988. *Ethics and Regulation of Clinical Research*. Second Edition. Yale University Press, CT.

Louis, K. S., Blumenthal, D., Gluck, M. E. and Soto, M. A. 1989. Entrepreneurs in academe: an exploration of behaviors among life scientists. *Administrative Science Quarterly* 34:110–131.

Lynch, C. 2008. Big data: how do your data grow? *Nature* 455:28–29.

Macrina, F. L. 2005. *Scientific Integrity*. Third Edition. American Society for Microbiology Press, Washington, DC.

Maher, B. 2010. Sabatage! *Nature* 467:516–518.

Malmgren, R., Ottino, J. and Nunes Amaral, L. 2010. The role of mentorship in protégé performance. *Nature* 465:622–626.

Mandavilla, A. 2011. Trial by twitter. *Nature* 467:286–287.

Martin, B. 1992. Scientific fraud and the power structure of science. *Prometheus* 10:83–98.

Martinson, B. C., Anderson, M. S. and De Vries R. 2005. Scientists behaving badly. *Nature* 435:737–738.

Maurer, H., Kappe, F. and Zaka, B. 2006. Plagiarism – a survey. *Journal of Universal Computer Science* **12**(8):1050–1084.

May, M. 2009. Sharing the wealth of data. *Scientific American Worldview* pp. 88–93.

McCook, A. 2009. Life after fraud. *The Scientist* **23**(7):28–33.

McGee, G. 2007. Me first. *The Scientist* **12**(9):28.

Munck, A. 1997. Examples of scientific misconduct. Pp. 31–33. Published in Elliott, D. and Stern J. E. (eds), *Research Ethics.* University Press of New England, Hanover, NH.

Nath, S. B., Marcus, S. C. and Druss, B. G. 2006. Retractions in the research literature: misconduct or mistakes? *Medical Journal of Australia* 185:152–154.

National Academy of Sciences. 2009. *Ensuring the Integrity, Accessibility, and Stewardship of Research Data in the Digital Age.* National Academies Press, Washington DC.

National Center for Biotechnology Information. 2008. GenBank Statistics: Growth of GenBank. Retrieved 2010: http://www.ncbi.nlm.nih.gov/genbank/genbankstats.html.

National Institutes of Health. 2009. Research and training grants: Success rates by mechanism and selected activity codes. Retrieved 2010: http://report.nih.gov/nihdatabook/.

Nelson, B. 2009. Empty archives. *Nature* 461:160–163.

Normile, D. 2009. *Science* retracts discredited paper; bitter patent dispute continues. *Science* 324:450–451.

Park, C. 2003. In other (people's) words: plagiarism by university students – literature and lessons. *Assessment and Evaluation in Higher Education* 28:471–488.

Pearson, H. 2009. Being Bob Langer. *Nature* 458:22–24.

Pecorari, D. 2003. Good and original: plagiarism and patchwriting in academic second-language writing. *Journal of Second Language Writing* 12:317–345.

Pennycook, A. 1996. Borrowing others' words: text, ownership, memory and plagiarism. *TESOL Quarterly* **30**(2):201–230.

Pool, R. 1997. More squabbling over unbelievable result. Pp.117–138. Published in Elliott, D. and Stern J. E. (eds), *Research Ethics.* University Press of New England, Hanover, N.H.

Prasher, D. C., Eckenrode, V. K., Ward, W.W., Prendergrast, F. G. and Cormier, M. J. 1992. Primary structure of the *Aequorea victoria* green fluorescent protein. *Gene* 111:229–233.

Pryor, E. R., Habermann, B. and Broome, M. E. 2007. Scientific misconduct from the perspective of research coordinators: a national survey. *Journal of Medical Ethics* 33:365–369.

Qiu, J. 2010. Publish or perish in China. *Nature* 463:142–143.

Refinetti, R. 2011. Publish and flourish. *Science* 331:29.

Reich, E. S. 2009. *Plastic Fantastic: How the Biggest Fraud in Physics Shook the Scientific World.* Palgrave Macmillan, NY.

Reich, E. S. 2010. Self-plagiarism case prompts calls for agencies to tighten rules. *Nature* 468:745.

Reis, R. 1999. Choosing a research topic. *The Chronicle of Higher Education.* Published Online: http://chronicle.com/article/Choosing-a-Research-Topic/45641/.

Rosser, M. and Yamada, K. M. 2004. What's in a picture? The temptation of image manipulation. *Journal of Cell Biology* 166:11–15.

Scollon, R. 1995. Plagiarism and ideology: identity in intercultural discourse. *Language in Society*. 24:1–28.

Shamoo, A. E. and Resnik, D. B. 2003. *Responsible Conduct of Research.* Oxford University Press, NY.

Siegel, D. and Baveye, P. 2010. Battling the paper glut. *Science* 329:1466.

Smith, R. 2006. Research misconduct: the poisoning of the well. *Journal of the Royal Society of Medicine* 99:232–237.

Spier, R. 2002. The history of the peer-review process. *Trends in Biotechnology* 20:357–358.

Stewart, C.N. Jr. and Edwards, J. E. 2008. How to teach research ethics. *The Scientist* **22**(2):27.

Swazey, J., Anderson, M. and Louis, K. 1993. Ethical problems in academic research: A survey of doctoral candidates and faculty raises important questions about the ethical environment of graduate education and research. *American Scientist* 81:542–553.

Swazey, J. and Bird, S. 1997. Teaching and learning research ethics. Pp. 1–19. Published in Elliott, D. and Stern, J. E. (eds), *Research Ethics.* University Press of New England, Hanover, NH.

Thorisson, G. A. 2009. Accreditation and attribution in data sharing. *Nature Biotechnology* 27:984–985.

Titus, S., Wells, J. and Rhoades, L. 2008. Repairing research integrity. *Nature* 453:980–982.

Titus, S. and Bosch, X. 2010. Tie funding to research integrity. *Nature* 466:436–438.

Van Noorden, R. 2010. A profusion of measures. *Nature* 465:864–866.

Vasgird, D. and Fischbach, R. 2003. Who is an author? Retrieved 2009 from Responsible Conduct of Research, Columbia University: http://ccnmtl.columbia.edu/projects/rcr/rcr_authorship/case/index.html.

Waldrop, M. 2008. Wikiomics. *Nature* 455:22–25.

Wang, K. 2005. Education and penalties are key to tackling misconduct. *Nature* 436:626.

Warren, W. L. and Cao, S. A. 2009. Certainty not required for inventorship. *Genetic Engineering News* 29:10, 12.

Watland, A. M., Schauber, E. M. and A. Woolf. 2007. Translocation of swamp rabbits in eastern Illinois. *Southeastern Naturalist* 6:259–270.

Watson, J. D. 1968. *The Double Helix: a Personal Account of the Discovery of the Structure of DNA*. New American Library, NY.

Weil, V. and Arzbaecher, R. 1997. Relationships in laboratories and research communities. Pp.69–90. Published in Elliott, D. and Stern, J. E. (eds), *Research Ethics.* University Press of New England, Hanover, NH.

Weston, A. 2002. *A Practical Companion to Ethics*, Second Edition. Oxford University Press, NY.

Wilson, J. R. 2002. Responsible authorship and peer review. *Science and Engineering Ethics* 8:155–174.

Winter S. 2010. Former Wisconsin researcher sentenced for misconduct. BioTechniques Newsletter September 23 2010, http://www.biotechniques.com/news/Former-Wisconsin-researcher-sentenced-for-misconduct/biotechniques-302891.html?utm_source=BioTechniques+Newsletters+%2526+e-Alerts&utm_campaign=d9d482b24e-BioTechniques_Weekly&utm_medium=email (last accessed November 17 2010).

Woolf, P. 1997. Pressure to publish and fraud in science. Pp. 141–145 Published in Elliott, D. and Stern, J. E. (eds), *Research Ethics.* University Press of New England, Hanover, NH.

Wuchty, S., Jones, B. F. and Uzzi, B. 2007. The increasing dominance of teams in production of knowledge. *Science* 316:1036–1039.

Yank, V. and Barnes, D. 2003. Consensus and contention regarding redundant publications in clinical research: cross-sectional survey of editors and authors. *Journal of Medical Ethics* 29:109–114.

Zhang, Y. 2010. Chinese journal finds 31% of submissions plagiarized. *Nature* 467:153.

Index